Peptide Synthesis

Edited by Jaya T. Varkey

Published in London, United Kingdom

IntechOpen

Supporting open minds since 2005

Peptide Synthesis
http://dx.doi.org/10.5772/intechopen.79243
Edited by Jaya T. Varkey

Contributors
Ali Nabipour Chakoli, Tayfun Acar, Burcu Ucar, Pelin Pelit Arayici, Melis Sen, Zeynep Mustafaeva, Serap Derman, S Silpa, Yoshio Hamada, Artemis Stamboulis, Marc Bruggeman, Hanieh Ijakipour, Jaya T Varkey

First published in London, United Kingdom, 2019 by IntechOpen
IntechOpen is the global imprint of INTECHOPEN LIMITED, registered in England and Wales, registration number: 11086078, 7th floor, 10 Lower Thames Street, London, EC3R 6AF, United Kingdom
Printed in Croatia

British Library Cataloguing-in-Publication Data
A catalogue record for this book is available from the British Library

Additional hard and PDF copies can be obtained from orders@intechopen.com

Peptide Synthesis
Edited by Jaya T. Varkey
p. cm.
Print ISBN 978-1-83968-257-5
Online ISBN 978-1-83968-258-2
eBook (PDF) ISBN 978-1-83968-259-9

For EU product safety concerns:
IN TECH d.o.o., Prolaz Marije Krucifikse Kozulić 3, 51000 Rijeka, Croatia, info@intechopen.com or visit our website at intechopen.com.

Meet the editor

Jaya T. Varkey, PhD, graduated with a degree in Chemistry from Cochin University of Science and Technology, Kerala, India. She obtained a PhD in Chemistry from the School of Chemical Sciences, Mahatma Gandhi University, Kerala, India, and completed a post-doctoral fellowship at the University of Minnesota, USA. She is a research guide at Mahatma Gandhi University and Associate Professor in Chemistry, St. Teresa's College, Kochi, Kerala, India. Dr. Varkey received a National Young Scientist award from the Indian Science Congress (1995), a UGC Research award (2016–2018), an Indian National Science Academy (INSA) Visiting Scientist award (2018–2019), and a Best Innovative Faculty award from the All India Association for Christian Higher Education (AIACHE) (2019). She Hashas received the Sr. Mary Cecil prize for best research paper three times. She was also awarded a start-up to develop a tea bag water filter.
Dr. Varkey has published two international books and twenty-seven international journal publications. She is an editorial board member for five international journals.

Contents

Preface

Peptide synthesis includes an array of techniques and procedures that enable the preparation of materials ranging from small peptides to large proteins. The pioneering work of Bruce Merrifield, which introduced solid phase peptide synthesis (SPPS), dramatically changed the strategy of peptide synthesis. But efficient synthesis of long chain and difficult peptide sequences is still troublesome because of the tendency to aggregation. Moreover low solubility of these peptides makes synthesis and purification laborious. This book gives a detailed review on synthesis and applications of synthetic peptides. Various aspects of solid phase peptide synthesis including different resins, linkers and synthesis techniques are discussed in detail. A novel method utilizing '*O*-isoacylpeptide' for the effective synthesis of long and difficult peptide sequences is presented. A section is devoted to a discussion of antimicrobial peptides, especially on the activity of defensin-like peptides.

This book is intended for those researchers who wish to study the synthesis of long chain and difficult peptides or who wish to become actively involved in the production of long chain peptides. As such it strives to provide practical information on the common problems relating to many aspects of synthetic peptides. At the same time this book is current and comprehensive and will be of general interest to peptide chemists.

Jaya T. Varkey, PhD
St. Teresa's College,
Kochi, Kerala, India

Chapter 1

Introductory Chapter: Peptide Synthesis

Jaya T. Varkey

1. Introduction

Peptide science is presently witnessing an enormous growth in its synthetic developments. Many synthetic peptides have commercial and pharmaceutical applications. But the synthesis of these peptides is a difficult task. The advent of solid-phase peptide synthesis has changed the face of peptide synthesis by opening the way to the extensive use of synthetic peptides in chemical and biomedical applications. Along with this landmark discovery, many other developments such as improved synthetic strategies, selection of protecting groups, automated synthetic methods, and advanced purification and analytical techniques were also evolved [1]. This book is intended essentially for those investigators who wish to make use of synthetic and antimicrobial peptides in their research and also provide practical information regarding the synthesis of difficult sequence-containing peptides. At the same time, it is addressing the common problems relating the synthesis and applications of synthetic peptides.

2. Methods of peptide synthesis

Methods for the chemical synthesis of peptides are divided into two groups: classical solution phase and solid-phase peptide synthesis (SPPS). Solution phase is the traditional way for large-scale synthesis of peptides. But these methods are labor-intensive and time-consuming because of its intermediate purification procedures and unforeseeable solubility issues. Hence many researchers who need to synthesize peptides choose the more convenient solid-phase approach. One of the main difficulties in solid-phase assembly of peptides is that of obtaining reasonable quantities of pure peptides. Investigations dealing with the quantitative aspects of polymer-supported reactions have shown that the insoluble support does have a significant dynamic influence on the bound substrates. An efficient polymeric support for peptide synthesis should have optimum hydrophobic-hydrophilic balance compatible with the peptide being synthesized. Systematic studies on the polymer-supported reactions show that the use of a flexible polymer support enhances the reactivity [2]. But the design and development of polymer supports having appropriate hydrophobic-hydrophilic balance is a difficult task.

3. *O*-isoacylpeptide method

Various methods are reported for the efficient synthesis of long-chain and difficult peptide sequences. These peptides have a tendency of aggregation and low solubility, making its synthesis and purification laborious. Native chemical

 IntechOpen

ligation (NCL) renders a highly effective and powerful method for the preparation of long-chain peptides [3]. This book introduces a novel "*O*-isoacylpeptide" method for the effective synthesis of long and difficult peptide sequences as in [4]. *O*-isoacylpeptides having ester bonds can be converted into the parent peptides under physiological condition *via* the *O* to *N* intramolecular migration. In addition a new approach called segment condensation with no racemization using the *O*-isoacylpeptide method is also introduced excellently [5]. This segment condensation can be treated as a substitute for NCL method.

4. Synthetic peptides

This book gives a detailed review on synthesis and applications of synthetic peptides. Various aspects of solid-phase peptide synthesis including different resins, linkers, and synthesis techniques are discussed in detail. Applications of synthetic peptides as peptide vaccines, radio-theranostic agents, radio-labeled peptide analogues, and radio peptides for imaging therapy are featured as part of the synthetic peptide chapter. This chapter also presents an excellent discussion on cell-penetrating peptides (CPP) [6] as molecular carriers. It not only illustrates the use and applications of cell-penetrating peptides but also provides different aspects of various cell-penetrating techniques. A more in-depth description of factors affecting the mechanism of cellular uptake and various molecular detections of CPPs are also included in this chapter.

5. Antimicrobial peptides

Antimicrobial peptides are a growing group of natural and synthetic peptides with a broad spectrum of targets including viruses, bacteria, fungi, and parasites [7]. Another interesting feature of this book is a comprehensive discussion on antimicrobial activity of defensin-like peptides [8]. These antimicrobial peptides act mainly by damaging the bacterial cell membranes and are found in many parts of the body. The chapter on defensins presents a discussion on mammalian defensins, their antimicrobial mechanisms, and its various evaluation techniques. This section also compares the antimicrobial activity of defensins in free form and immobilized on material surfaces along with pictorial representation of various immobilization methods. Furthermore, several of the concepts discussed can be easily adopted for the next-generation antimicrobial surfaces as coatings for medical devices and implants.

Peptide synthesis has come a long way from the invention of solid-phase synthesis to the present day. Each chapter in this book portrays one major aspect of peptides including its synthesis and applications. This book is intended to provide the researchers with various new methods for the synthesis of long and difficult sequence peptides.

Author details

Jaya T. Varkey
Department of Chemistry, St. Teresa's College, Ernakulam, Kochi, India

*Address all correspondence to: jayavarkey@yahoo.com

References

[1] Grant GA. Synthetic Peptides: A User's Guide. New York: Oxford University Press; 2002

[2] Varkey JT, Rajasekharan Pillai VN. Merrifield resin and newly developed 1,6-hexanediol diacrylate resin for solid phase peptide synthesis—A comparative study. Journal of Applied Polymer Science. 1999;**71**:1933-1939

[3] Johnson ECB, Kent SBH. Insights into the mechanism and catalysis of the native chemical ligation reaction. Journal of the American Chemical Society. 2006;**128**:6640-6646

[4] Hamada Y, Matsumoto H, Yamaguchi S, et al. Water-soluble prodrugs of dipeptide HIV protease inhibitors based on *O*→*N* intramolecular acyl migration: Design, synthesis and kinetic study. Bioorganic & Medicinal Chemistry. 2004;**12**:159-170

[5] Hamada Y, Kiso Y. New directions for protease inhibitors directed drug discovery. Biopolymers. 2016;**106**:563-579

[6] Copolovici DM, Langel K, Eriste E, Langel U. Cell-penetrating peptides: Design, synthesis, and applications. ACS Nano. 2014;**8**(3):1972-1994

[7] Epand RM, Vogel HJ. Diversity of antimicrobial peptides and their mechanisms of action. Biochimica et Biophysica Acta (BBA) - Biomembranes. 1999;**1462**(1-2):11-28

[8] Wilson SS, Wiens ME, Smith JG. Antiviral mechanisms of human defensins. Journal of Molecular Biology. 2013;**425**(24):4965-4980

Chapter 2

Synthesis and Applications of Synthetic Peptides

Burcu Ucar, Tayfun Acar, Pelin Pelit Arayici, Melis Sen, Serap Derman and Zeynep Mustafaeva

Abstract

The synthesis and applications of the peptides are gaining increasing popularity as a result of the developments in biotechnology and bioengineering areas and for a number of research purposes including cancer diagnosis and treatment, antibiotic drug development, epitope mapping, production of antibodies, and vaccine design. The use of synthetic peptides approved by the health authorities for vaccine, for cancer, and in drug delivery systems is increasing with these developments. The aim of this book chapter is to review the recent developments in the use of peptides in the diagnosis of drug and vaccine systems and to present them to the reader with commercially available illustrations.

Keywords: synthetic peptide, solid-phase peptide synthesis (SSPS), peptide therapeutics, peptide vaccines, cell-penetrating peptides

1. Introduction

The aim of this chapter is to review some applications of synthetic peptides providing a brief knowledge about peptide synthesis. In the first part, information about the peptide synthesis was given in a very simple and readable format under the title of solid-phase peptide synthesis including a brief history, solid supports, linkers, protecting groups, and analysis method sections. Then the synthetic peptide vaccine application of peptides was reviewed. After that, the topic of nuclear imaging-guided peptidic drug targets and labeling techniques and recent developments in therapy was discussed. In the last part, information about cell-penetrating peptides that can be used as molecular carries is mentioned with providing classification and cellular uptake mechanism of them.

2. Solid-phase peptide synthesis (SSPS)

The specific characters of peptides (high bioactivity, high specificity, and low toxicity) have made them attractive therapeutic agents. The synthesis of the peptides may provide sufficient material to enable further studies and to determine the structure-activity relationships or may provide discovery of new analogues with improved properties [1–7]. The peptides are able to synthesize in three methods: in a solution medium, on a solid support, or as a combination of the solid and the solution synthesis. Although peptide synthesis is often carried out by the solid-phase method, the solution method was preferred by the pharmaceutical companies in the 1970s and 1980s [8]. In the solution medium synthesis method, except for the reversible protection of the N-amino group of the first amino acid or fragment, the orthogonal protection of carboxyl groups of the second amino acid or the fragment is needed. On the solid-phase method, the synthesis is carried out on a solid support, also called a resin. The peptide is separated from the resin after each amino acid in the peptide sequence is sequentially bound. In the solid-phase technique, the peptide that bounded to the insoluble resin is separated without any significant loss during the washing or filtration of the resin. All reactions are carried out in a single reaction vessel, and possible losses are prevented during processes such as exchange and transfer of reaction vessels [9]. Another one is hybrid synthesis which is the composition of the solution and SPPS methods. Herein, the peptide to be synthesized is obtained after the condensation from a solution of two or more suitable peptide sequences, which are obtained mainly by solid-phase synthesis [10].

The principle of peptide synthesis in the solid phase is quite simple. The peptide chain is attached to the stable solid phase. The peptide sequence remains bound on this resin for the duration of the synthesis. During the synthesis that includes deprotection, activation, and coupling steps, other soluble chemicals outside the resin beads are removed by filtration and washing. In the last stage, the desired product is separated from the solid phase by cleavage procedure. Purification and characterization procedures are carried out in the free solution of the desired product [11]. In SPPS, although there are two main strategies, the Boc approach requires the use of hard acids such as HF; thus the majority of synthetic peptides are nowadays synthesized by using Fmoc chemistry. The most important advantage of the Fmoc method compared to Boc strategy is that it does not require corrosive acids such as TFA during the synthesis cycle, and thus the synthesis can be automated. In this method, the N-terminal amine groups, the acid-labile side-chain protecting groups, and the linkers forming the protecting group of the C-terminal amino acid are protected by the base-labile Fmoc group [12, 13]. Solid support materials (resins), linkers, protecting groups, and cleavage cocktail solutions are general components of the solid-phase peptide synthesis.

The resin is a spherical polymer ball bearing the active groups to which the first amino acid (indicates the first amino acid at the C-terminus of the sequence) of the sequence to be synthesized can be attached. The solid supports used in the solid-phase peptide synthesis must be stable against mechanical mixing, various temperature conditions, and different types of solvents, must have a narrow range of resin bead sizes, and must have high swelling properties (should be able to swell up to 5–6 times in DCM, dimethylformamide (DMF), etc.); thus the reagents can reach the active regions of the resin. The solid support material is usually used in a size of about 100–200 mesh or 200–400 mesh, and the peptide chain is extended by the addition of amino acids on each other on the resin. The most common classes of resins are the classic polystyrene (PS) resins (cross-linked with 1% of divinylbenzene (DVB)), the PS-functionalized polyethylene glycol (PEG) resins, pure cross-linked PEG resins, and polystyrene cross-linked with 1,6-hexanediol diacrylate resin (PS-HDODA) [14–20]. The solid support used in the synthesis can be of two kinds depending on whether or not the initial amino acid of the sequence to be synthesized is bound to the resin. Accordingly, there are two types of resin, either pre-loaded or unloaded, but pre-loaded resin is generally preferred. The pre-loaded resins differ depending on whether they are lowly loaded or highly loaded. When the highly loaded resin is used, aggregation of the peptide sequences growing on the resin increases, and this will lead to the deletion of an amino acid which is an undesirable situation; therefore, the use of lowly loaded resins is common [19].

In order to bind the first amino acid to the resins, chemical structures containing functional group(s) called *linker* are attached. The number of these active groups on the resin (substitution) is important in calculating the theoretical yield of a peptide to be synthesized and in determining the amount of chemicals (deprotection, activation, coupling) required for the synthesis of a peptide. Depending on the functional group of the C-terminal end of the target peptide sequence, peptide acids or peptide amides can be synthesized by means of an ester or an amide bond, respectively. While the peptide acids are obtained using 4-benzyloxybenzyl alcohol (Wang), 2-(4-hydroxymethyl) phenoxyacetic acid (HMPA), superacid-sensitive resin (SASRIN), and 2-chlorotrityl linkers, the peptide amides rink amide, 4-methyl benzhydryl amide (MBHA), and Sieber and primary amide (PAL) linkers are required (**Figure 1**) [13]. After the cleavage procedure, C-terminus of the peptide sequence remains —COOH or —NH_2 depending on the type of the linker on the resin beads [13, 19, 21, 22].

The reaction between the activated carboxylic group of one amino acid and the amino group of another amino acid is needed for a peptide linkage. In order to prevent side reactions from occurring, the remaining functional groups of each amino acid must be appropriately protected. For this purpose, many *protecting groups* are used to prevent the reaction of different types of functional groups. Because each amino acid contains different groups in its side chains, the protecting groups which prevent these groups from reacting vary according to the amino acid and used methodology (Fmoc or Boc chemistry). Some amino acids have not any functional groups to react in the side groups; only alpha-amino groups of these amino acids are protected by Fmoc or Boc [23–33].

Cleavage is one of the most important steps in solid-phase peptide synthesis. Exposing a resin to a cleavage cocktail is not an easy reaction; it brings a series of competitive reactions. Unless appropriate reagents and reaction conditions are selected in the cleavage process, the desired peptide can be irreversibly modified or damaged. Also, since the DMF used in the solid-phase peptide synthesis can inhibit TFA acidolysis, before starting the cleavage, the peptide-bound resin should be washed with DCM to remove all DMF. The goal of cleavage is to split the synthesized peptide sequence from the resin while also removing amino acid side-chain protecting groups. For this process, strong acids such as anhydrous HF, TFMSA, or TMSOTf are used in Boc chemistry, while cocktails containing TFA are preferred in the Fmoc process [12, 22, 34]. Additional substances such as EDT, phenol, and thioanisole called *scavengers* are put into the cleavage medium to prevent the cationic products occurring during the cleavage process to attack the amino acids that are having electron-rich side chains such as Trp, Tyr, and Met. Considering the amino acid types involved in the resin-linked peptide sequence synthesized by the solid-phase peptide synthesis method, it is decided to which type and how much scavenger add to the cleavage cocktail. For example, when there is one or more Trp in the peptide sequence, using EDT in the cleavage will substantially protect Trp against oxidation. Typically, the peptide-bound resin is treated using 95% TFA with gentle shaking for 1–3 h [12, 22, 34, 35].

The crude peptides synthesized by solid-phase peptide synthesis with the deprotection, activation, and coupling and cleavage steps are mostly analyzed on analytical HPLC using C18 columns at a concentration of 1 mg/mL with gradient elution method of water (0.1% TFA) and acetonitrile (0.1% TFA) as mobile phase. If purification of the synthesized peptide is required according to the application to be used, preparative HPLC is used. For the analysis, the peptide should be dissolved in a minimum volume of 0.1% TFA in water, and the acetonitrile and water gradient elution is adjusted according to the polarity of the peptide [12, 22, 35–37]. The molecular weight of the synthesized peptide is confirmed by mass spectroscopy. The most commonly used methods are LC-MS, MALDI-TOF, and LC-QTOF-MS. Systems with such a combination of chromatography and spectroscopy are useful for the complete characterization of synthetic peptides. While chromatographic analyses allow us to have knowledge about the peptides' purity, the molecular weight of the peptides is determined by MS analyses. It is also found in some systems that determine the amino acid sequence of the peptide molecule like amino acid analyzer and LC-MS-MS. Information on structural properties of peptides can also be obtained by NMR and FTIR methods. With NMR and FTIR, the structures of the peptides are further elucidated via specific binding and functional groups [12, 35]. Moreover circular dichroism (CD) spectroscopy gives information about the conformation and secondary structures of polypeptides. This technique can be used to distinguish between random coil, alpha-helix, or beta-sheet structures [38, 39].

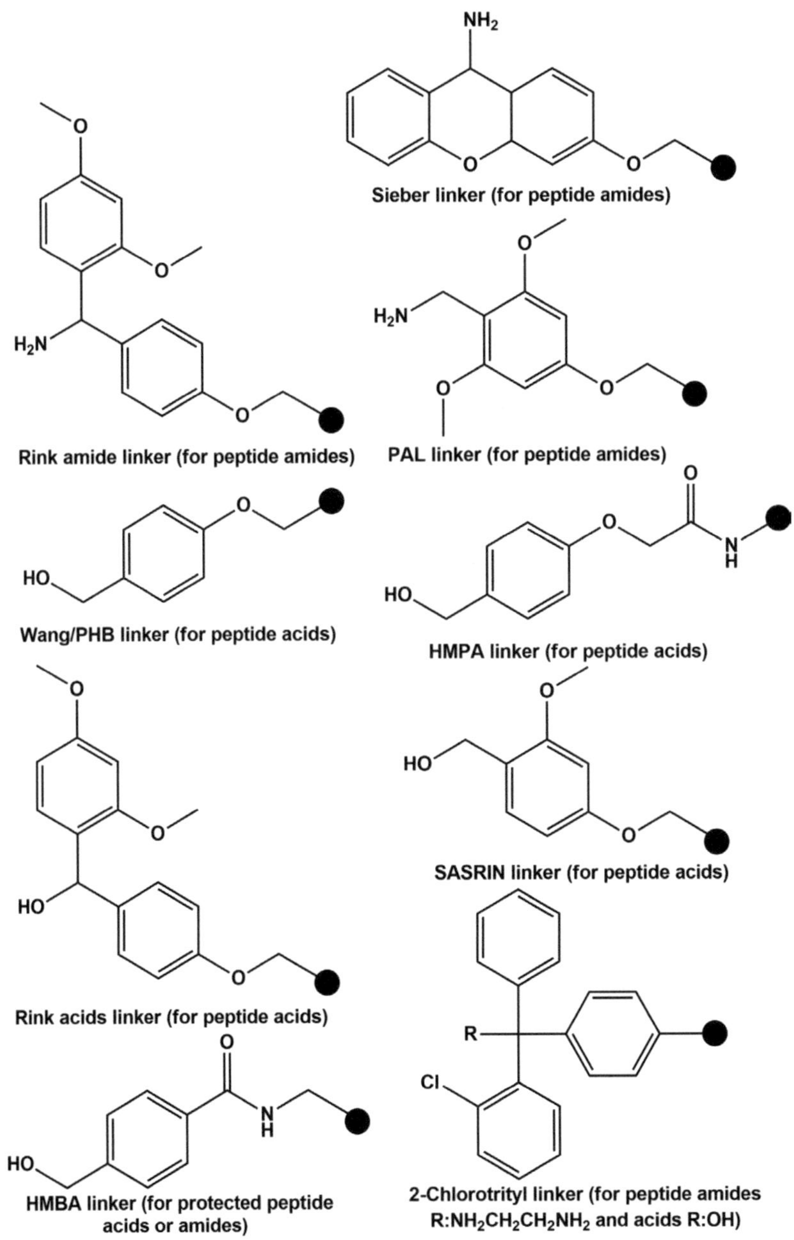

Figure 1.
Some of the fundamental linkers used on the Fmoc-based SPSS [22].

3. Peptide vaccines

The increase in studies on peptide vaccines in recent years shows that peptide vaccines will be an important part of new-generation vaccine systems. Vaccines are indispensable in protecting human and community health in terms of reducing infectious diseases, disability, and deaths and, most importantly, eliminating and eradicating the long-term disease. The development of the vaccine, which began with the observations of Edward Jenner at the end of the eighteenth century, has significantly reduced the number of infections and diseases until now [40–42]. Traditional vaccines (live-attenuated or inactivated vaccines) have been widely used over the last hundred years to develop effective vaccines against many diseases. However, the use of this technology is a source of concern because of the problems encountered in the preparation and use of such vaccines. The disadvantages of classical vaccines such as the increased need for safety, high cost in the production of large-scale vaccine preparations, high genetic variability of pathogens, side effects, and risk of leak of the disease agent have necessitated research to develop new technological vaccines [43, 44]. The basis for new vaccine technologies is the production of high-purity subunit vaccines that contain a small part of the pathogen necessary to generate an immune response. Subunit vaccines can be composed of polysaccharides and proteins or their peptide fragments. Synthetic peptide vaccines consist of 20–30 amino acids containing the specific epitope of a corresponding antigen against various diseases. There is no limitation from cancer diseases to allergies in diseases targeted by peptide vaccines. The advantages of peptide vaccines have enabled peptides to be preferred in vaccine technology [45–48]. Although many peptide vaccine studies continue, there is no approved peptide vaccine for human use. According to the database of clinical studies maintained on ClinicalTrials.gov, in the search on the topic of peptide vaccine until mid-December 2018, 374 clinical trials in phase I, 272 clinical trials in phase II, 14 clinical trials in phase III, and 2 clinical trials in phase IV have been found [47, 49]. Furthermore, it is reported in the literature that a veterinary peptide vaccine against canine parvovirus, which is targeted against animal diseases, is effective [50]. Advantages involved in the choice of peptide-based vaccine can be summarized as follows: Peptide-based vaccines are more reliable than classical vaccines because they are produced using chemical synthetic approaches almost exclusively. It does not produce any unwanted immune response as it does not contain biological material. Also, this situation eliminates the risk of biological contamination. The rapid, easy, inexpensive, and high-purity synthesis of peptides by microwave-assisted solid-phase peptide synthesis methods is an important advantage for the use of peptides in vaccine systems. These synthesized peptide antigens can be fully characterized by analytical methods. "Cold chain" is not required generally during transport and storage of peptide vaccines. Peptide vaccines have high stability and are typically water-soluble. Allergic and/or reactogenic effects are avoided by the usage of peptide vaccines. Peptide vaccines can be customized using the peptide epitope of one or more antigens [46, 51, 52].

Despite all its advantages, there are some limitations of peptide-based vaccines. The peptides alone are generally weakly immunogenic, have poor release properties and are easily undergo enzymatic degradation. These disadvantages can be overcome using adjuvants and delivery systems. Adjuvants, derived from the Latin word "adjuvare," which means "to help," play a crucial role to the elicitation of a strong immune response to an antigen [49, 53–56]. The first step in the development of peptide vaccines is the identification of the structure of peptide epitopes which will provide effective and long-term immunity. After the synthesis and characterization of specific antigens to be determined, the appropriate adjuvant and delivery system selection is carried out [43, 46]. Currently, instead of classical adjuvants, the use of new generation of more effective adjuvants is being studied. It is aimed to overcome the disadvantages of the traditional adjuvants with the development of modern adjuvants and delivery systems. Alum (insoluble salts of aluminum), which is a classical adjuvant, is a weak immune stimulant; its mechanisms of action are not fully elucidated and have a possibility of side effects. Emulsion adjuvants such as incomplete Freund's adjuvant (IFA), complete Freund's adjuvant (CFA), and lipid A are used; however, it is important to develop safer adjuvants due to their toxicity [46, 49, 57, 58]. Protein and synthetic polymer carriers play an important role in new-generation vaccine systems. The immunogenicity of peptide vaccines based on conjugation to protein and polymeric carriers is enhanced. Bovine serum albumin (BSA), keyhole limpet hemocyanin (KLH), human serum albumin (HSA), and ovalbumin (OVA) are the most preferred proteins in peptide-protein conjugates [51, 59, 60]. In recent years, peptide-polymer conjugations using synthetic polymeric carriers have gained importance. Polymers to be selected in peptide-polymer conjugates, should be biocompatible, be relatively easy to synthesize and be modifiable according to the recommended use. This situation is pointed out that such polymers are important carrier candidates for peptide vaccines [61–63].

Nanoparticles are investigated for their potential use as vaccine delivery systems. The use of nanoparticles in peptide-based vaccines may be useful in elimination of limitations with small peptide antigens. It can play a critical role in increasing the size of the antigen molecule, protecting the peptides from enzymatic degradation, and the delivery of the peptides to the targeted cells. Inorganic NPs, lipid-based NPs, polymeric NPs, and carbohydrates are common classes of nanoparticles. Especially polymeric NPs are of great interest for vaccination due to their biocompatibility and predictable degradation. In addition to natural polymers such as chitosan, dextran, and albumin, synthetic polymers such as polyesters, polyanhydrides, and polyamides can be mentioned. Mostly studied poly(lactic-co-glycolic acid) (PLGA) is a synthetic copolymer composed of lactide and glycolide and approved by the European Medicine Agency (EMA) and US Food and Drug Administration (FDA) [64]. This polymer, which is biodegradable and biocompatible, is suitable for vaccine and drug delivery systems in humans. In the usage of polymeric nanoparticles as delivery systems for peptide vaccines, the peptide may be covalently/non-covalently attached to the surface of the particles through functional groups, or the peptide may be loaded into the nanoparticles [65–69].

Peptide vaccines are considered to be critical in the production of effective, safe, inexpensive, and easy to produce vaccines by the abovementioned advantages and the development of new adjuvants.

4. Utilizing of peptides as radio-theranostic agents

This part of the chapter will introduce the chemistry behind the radiolabeling of peptide-based diagnostic and/or therapeutic pharmaceutical systems used for nuclear oncological imaging and present research related to drug discovery in various areas such as chemistry, bioengineering, nanoscience, and nuclear oncology [70]. Synthetic peptide-based drugs labeled by radiochemically different techniques are used in routine clinical applications for diagnosis and therapy of diseases. Peptides targeting the somatostatin receptor subtypes have been routinely applied for peptide receptor radionuclide therapy (PRRT) of neuroendocrine and prostate cancers. Tumor regression, lengthening survival, and increased quality of life were observed in the patients treated with 177Lu-/90Y-labeled peptide conjugates [71]. This section provides an overview of various targeting and labeling techniques for receptor-targeted imaging.

4.1 Radiolabeled peptide analogues

PRRT with radiolabeled peptide pharmaceuticals is a suitable and promising next-generation therapy method for inoperable patients with metastatic tumor. Radiolabeled peptide ligands are effective vectors for both detection and treatment of tumor cells overexpressing receptors specifically binding these ligands. Such radiopharmaceuticals, called as "theranostics," can be administered for diagnosis besides treatment, depending on the type of radionuclide being bound to the peptide sequence. Especially somatostatin and prostate-specific membrane antigen (PSMA)-based peptide analogues have significantly influenced the staging and therapy of patients. Mentioned analogues are used in clinical routine as an integral part of the treatment [72].

4.1.1 Radionuclide therapy using somatostatin analogues

PRRT and radiolabeled somatostatin (SST) peptide analogues are highly effective treatment agents. Somatostatin is in the form of dominant but functionally less active 14 amino acids called SST-14 and SST-28 in the form of a larger and stronger 28 amino acids [73]. According to the structure-activity researches of SST-14, although the amino acid residues of Phe, Trp, Lys, and Thr which contain a β-turn are important for bioactivity, different amino acids may be replaced by the Phe and Thr amino acids for this bioactivity, whereas Trp and Lys residues are definitely required. Octreotide, lanreotide, vapreotide, and pasireotide, which are the 4SS analogue synthetic peptides, are currently used actively in routine clinical applications. The octreotide compound, which is the synthetic peptide compared to the natural SST, was found to be 19 times more effective in inhibition of growth hormone secretion [74]. The presence of the D-Phe at the N-terminal end and the L-Thr at the C-terminal end and the replacement of the L-Trp at position 8 with D-Trp ensure that the peptide is resistant to degradation [75]. The somatostatin receptor family contains five receptor subtypes, sst1–sst5. Most of the neuroendocrine tumors, especially subtype 2 (sst2), have a strong overexpression of sst. The clinical use of radiolabeled sst-targeting analogue [111In-DTPA^0]octreotide initiated the development of numerous somatostatin analogues. This analogue is used as a theranostic agent in positive metastatic tumors. Tumor regression and survival improvement can be provided with [90Y-DOTA^0,Tyr^3]octreotide and [177Lu-DOTA^0,Tyr^3]octreotate. Several phase-1 and phase-2 PRRT trials were carried out using [90Y-DOTA^0-Tyr^3]octreotide. Tumor regression responses in the clinical studies with these agents in patients who suffered from GEP-NETs ranged from 9 to 33% [70]. As a result of the administration of the radiolabeled antagonist 125I-JR11 and agonist 125I-Tyr^3-octreotide in different tumors, it was found that the antagonist binds to the SSTR2 region at a rate of 3.8–21.8-fold higher [76]. This significantly increased binding not only provides a more accurate localization in tumor and metastasis but also enables therapeutic interventions with radiolabeled SSTR antagonists to be more effective. Despite the low SSTR2 concentrations, tumors other than GEP-NETs and lung NETs have been targeted with SSTR2 antagonists; tumors of these types are breast cancer, medullary thyroid cancer, non-Hodgkin lymphomas, renal cell cancer, and small-cell lung cancer [76, 77].

4.1.2 Radionuclide therapy using PSMA analogues

PSMA has become an interesting target for the diagnosis and therapy of peptide receptors for small molecule ligands [78]. Since PSMA is often overexpressed in prostate cancer (PCa) cases, many different PSMA ligands are aimed to be targeted by the same biological mechanism to diagnose and treat metastatic castration-resistant prostate cancer (mCRPC). Many studies proved the superiority of 68Ga-PSMA PET/CT (positron emission tomography/computed tomography) as compared to CT, MRI, or bone scan for determination of metastases for first staging at initial diagnosis [79, 80]. Imaging with PSMA in nuclear medicine substantially affected the detection and treatment of patients with prostate cancer. PSMA has been known to be promising and frequently preferred in advanced clinical studies because of providing preliminary information for different types of clinical conditions and detecting lesions with low PSA levels [81]. PET/CT data obtained using PSMA provides a noninvasive evaluation of PSMA expression and is used for restaging prostate cancer after radical prostatectomy when PSA level is low [82]. Also, it can be applied in endoradiotherapy because of the intracellular internalization feature of PSMA. A small molecule inhibitor of PSMA ((S)-2-(3-((S)-1-carboxy-5-(3-(4-[124I]iodophenyl) ureido) pentyl)ureido)pentanedioicacid; MIP-1095) has been administrated to evaluate potential therapeutic use of 131I-MIP-1095 in men with mCRPC [83]. Due to their molecular and specificity similarity, 68Ga-PSMA-11, 68Ga-PSMA-617, and 68Ga-PSMA-I&T are abbreviated as 68Ga-PSMA. 68Ga-PSMA-11 biodistribution was known to correspond well to cellular detection of PSMA expression. 68Ga-PSMA-617 and 68Ga-PSMA-I&T peptidic ligands demonstrated similar distribution and imaging properties [79].

4.1.3 Radionuclide therapy using other peptide analogues

For clinical applications, it is preferred that the peptide analogues are metabolically stable. The natural structure of the peptides makes them sensitive to peptidase and rapidly breaks down in blood and other tissues, so their potential uses as radiopharmaceuticals are restricted. Strategies for stabilizing peptides include incorporating biologically insoluble peptide bonds, stabilized amino acid derivatives, and cyclization [70]. Although the stability is advantageous, it is necessary to keep the nuclide in the tumor area for a long period of time for a good target-to-nontarget ratios and to rapidly clear the nuclide from nontarget tissues and blood. Researches on the topic of peptide-based radiopharmaceuticals have mostly been focused on receptor agonists. However, in recent years, somatostatin and bombesin antagonist peptide analogues have also been best shown for receptor targeting [72].

4.1.3.1 GRP receptor-targeting peptides

Overexpression of gastrin-releasing peptide receptors (GRPR) is seen in many human tumors, including prostate and breast tumors. Bombesin (BN), a 14-amino acid peptide with high affinity to the GRP receptor, is used for GRP receptor-targeted tumor imaging and treatment. 111In- and $^{99}Tc^{m}$-labeled BN analogues have been improved for SPECT imaging of GRP receptors. After a while, a novel DTPA-bound BN analogue, [111In-DTPA-ACMpip5, Tha6, βAla11, Tha13, Nle14] BN (5–14) (Cmp 3), was synthesized. Replacement of the DTPA chelator in the 111In-Cmp3 analogue with a DOTA makes the compound suitable for therapeutic use and diagnostic PET imaging. Another promising peptide-conjugated DOTA is 177Lu-AMBA (DOTA-BN (7–14)), which exhibits good tumor-background ratios. The 177Lu-AMBA is excreted through the kidneys, but kidney excretion cannot be reduced by co-lysine injection because there is not any lysine in the peptide structure. Nowadays, it is the first choice in the targeting of 68Ga GRP receptors as a positron emitter radionuclide which provides shorter half-life and in-house radiolabeling procedures as well as more accurate high-resolution PET images. Recently, imaging studies have been studied to detect a GRPR and integrin αvβ3 dual targeting tracer 68Ga-BBN-RGD for PET/CT imaging of metastatic breast cancer [84]. Recent researches on BN analogues have focused on the DOTA chelating systems for multi-use options such as SPECT, PET, and PRRT [72].

4.1.3.2 NT receptor-targeting peptides

Neurotensin (NT) with 13 amino acids in its structure is expressed in the central nervous system, peripheral tissues, and gastrointestinal system [72]. Despite overexpression of neurotensin receptors in 75% of ductal pancreatic carcinomas, NT receptor expression of endocrine pancreatic tumors, pancreatitis, and normal pancreatic tissue is negative (somatostatin positive) [85]. In a clinical study using 111In-labeled DTPA and DOTA-conjugated NT analogues, a number of specific changes were made to unnatural amino acids, indicating that the C-terminus plays a key role in the binding and biological properties of the peptide, and there is not a critical change in receptor binding activity of the peptide [72]. Because of high renal involvement of 111In-NT analogues, PRRT is not considered appropriate with these analogues. With the discovery of three times more stable $^{99}Tc^{m}$-labeled NT (NT-XIX) analogue, a better tumor-to-kidney ratio, higher tumor involvement, and higher kidney excretion were achieved. $^{99}Tc^{m}$-Demotensin, which was a high-stability NT analogue, has a higher tumor-to-intestinal and tumor-to-liver ratios, but the most appropriate ratios for imaging pancreatic tumors were obtained with the 111In-labeled analogues [86].

4.1.3.3 CCK2 receptor-targeting peptides

It has been determined that in the majority of medullary thyroid carcinomas (MTCs), in a high rate of small-cell lung cancer patients, stromal ovarian cancers, astrocytomas, and some other tumor types have cholecystokinin-2 (CCK2) receptors. The CCK2 receptor-specific CCK peptide analogue was conjugated with DTPA for targeting this receptor [85]. The highest tumor uptake and too high renal involvement have been demonstrated at minigastrin analogues containing the CCK8 sequence. The addition of the histidine residues to the array almost reduces the kidney uptake by twofold. According to a study carried out in mice bearing the AR42J tumor, it was reported that the DOTA-conjugated HHEAYGWMDF peptide sequence exhibited the highest tumor-to-kidney ratio compared to all peptides studied [86]. Additionally, 99Tc-labeled N4-derived analogues of minigastrin have been synthesized [72]. Although studies with these radioligands are still in the initial stage for PRRT, they have significance features for the future.

4.1.3.4 GLP-1 receptor-targeting peptides

The glucagon-like peptide 1 (GLP-1) receptor which is overexpressed in human endocrine tumors, insulinomas, gastrinomas, and pheochromocytomas is a subtype of glucagon receptors used as a vehicle for in vivo tumor targeting. Because natural GLP-1 receptor agonists are metabolized quickly in the blood, a more stable and specific exendin peptide has been developed for use in the scintigraphic imaging of GLP-1 receptor-expressing tumors [72]. In the following years, 111In-DTPALys40-exendin-4 [72] and [Lys40 (Ahx-DTPA-111In) NH_2] exendin-4 conjugates with high tumor-background ratios have been optimized with using animal models [87]. In a clinical study based on the development of GLP-1 receptor-targeting analogues, PET/CT imaging with the 68Ga-NOTA-exendin-4 peptide analogue has been reported to be a highly susceptible imaging technique for the detection of insulinoma localization [88].

4.1.3.5 *αvβ3 integrin-targeting peptides*

αvβ3 integrins are a transmembrane protein that can be expressed in proliferative endothelial cells and overexpressed in newly formed blood vessels where tumors are fed. The arginine-glycine-aspartic acid (RGD) tripeptide is essential for the interaction of extracellular matrix proteins to αvβ3 receptors. The cyclic RGD analogue containing these amino acids has the highest binding affinity. Many radiolabeled DTPA and DOTA-RGD conjugates with 111In, 90Y, 177Lu, 68Ga, and 64Cu which provide SPECT and PET imaging and PRRT have been discovered in recent years. Monomeric, dimeric, and tetrameric RGD peptides are bound to DOTA for developing receptor binding affinity and then radiolabeled with 111In. Although the monomeric and dimeric analogues have higher in vitro receptor affinity, the in vivo tumor uptake of the tetrameric analogue is higher. Also, it has been shown that multimeric RGD peptides are effective clinical molecules for in vivo determination of tumor angiogenesis in cancer patients [72].

4.2 Radiopeptides

In clinical practice, radiolabeled receptor ligands are used routinely for diagnostic imaging of overexpressed receptors and PRRT. Causes of clinical success of radiopeptidic receptor ligands are the following:

1. First, the presence of different radionuclides, having similar chemical properties, enables to label the same peptide with different radionuclides for different clinical purposes (SPECT, PET, PRRT).

2. Second, the influence of the high hydrophilic radiometal complex on peptide pharmacokinetics leads to rapid renal excretion and good target/background ratios.

3. Third, one-step in-house labeling methodology that facilitates the preparation of peptide radiopharmaceuticals in clinical routine [89].

The radioiodine which is used for radiolabeling of biologically active molecules is frequently preferred in PET imaging (124I), SPECT imaging (123/125I), treatment of different cancer types (131I), and biodistribution and kinetic investigation of novel peptide radiopharmaceuticals (125I). Direct radioiodination is based on the substitution of an aromatic proton with $*I^+$ (electrophilic radioiodide) and is successful only in electron-rich aromatic systems including activating substituents such as —OH, —NH_2, —OR, —NHCOR, or —SR. For in situ production of $*I^+$, chloramine-T (sodium tosylchloroamide) and iodogen (1, 3, 4, 6-tetrachloro-3a, 6a-diphenylglycoluril) are utilized. Chloramine-T is added to the reaction medium with sodium metabisulfite to terminate the labeling reaction and prevent oxidative damage. Oxidative enzymes (lactoperoxidase) are used for substitution of peptide sequences that have high sensitivity to oxidation. Tyr- and also His-containing peptides are readily radioiodinated in buffers such as phosphate buffered saline (PBS) or Tris (hydroxymethyl)-aminomethane (TRIS) at pH 7–8. Pre-radioiodination should be carried out by prosthetic groups. Selective prosthetic group conjugation is provided to the thiols with the help of pre-radioiodinated maleimides, and prelabeling of the corresponding peptide is carried out by the use of stannylated vinyl alkylating agents. In this way, tissue deiodinase and unfavorable structural conditions for radioiodination can be overcome [90].

Complex biomolecules such as peptides or proteins cannot be directly labeled with a highly basic [18F] fluoride by nucleophilic substitution and cannot tolerate labeling conditions. Activated aromatic precursors (NO_2, CN, CI, etc.) are substituents bound to the leaving group in the ortho- or para-position. Many receptor binding peptides such as octreotide, bombesin, neurotensin, and RGD analogues have been labeled using [18F]FP-NP(4-nitrophenyl-2-[18F]fluoropropionate) or [18F]SFB(N-succinimidyl-4-[18F]fluorobenzoate) [91]. Chemoselective strategies provide a one-step prosthetic group labeling reaction by unprotected precursors. Reaction of a 18F-labeled aldehyde with aminooxy- or hydrazino-functionalized peptides so-called click chemistry has recently found most popular application [89]. A suitable chelating agent is required for the radio metallization of the peptides. When a chelating agent is conjugated with a receptor binding peptide, it can affect both the binding affinity of the peptidic ligand and peptide pharmacokinetics [89]. Radiolabeling of peptides with the oxo-technetium ions (TcO_2^+,TcO^+) is carried out by using peptide-bound tetradentate $^{99}Tc^m$ chelators (N_3S or N_4 scaffold). The peptides are coordinated with donor groups such as amine, carboxylate, or hydroxyl of the HYNIC (hydrazinonicotinic acid) chelator. In order to initiate the labeling reaction, a generator eluate ($^{99}Tc^m$-saline) should be added into a vial containing all the mixtures. For the labeling with $^{99}Tc^m$, the $[^{99}Tc^m(CO)_3(H_2O)_3]^+$; disodium boronocarbonate, $Na_2[HBCO_2]$ (which also serves as an in situ CO source); and stannous ion, $SnCI_2$ (reducing agent) are used [89]. 68Gallium labeling reaction is initiated by eluting the 68Ge/68Ga generator using hydrochloric acid (HCl) (0.05–0.60 M). Approximately 2 mL of the eluent is transferred to the reaction vial. Reaction vial containing a mixture of the lyophilized DOTA-peptide conjugate and sodium acetate buffer with the eluent is heated for 15 minutes at 100°C (25 min at 90–95°C). The intermediate product is pushed through an extraction cartridge. The final product is analyzed to determine the labeling efficacy and purity [92].

5. Cell-penetrating peptides as molecular carries

The selective permeability and hydrophobic profile of the cellular membranes provide strict control of the molecular changes between the cytosol and the extracellular environment [93–95]. Generally, peptides are selective and effective signaling molecules which bind to specific cell surface receptors that are involved in physiological mechanisms such as peptides, hormones, neurotransmitters, growth factors, G-cell receptors (GPCRs), and ion channel ligands [96]. This characteristic of the peptides mentioned above and their attractive pharmacological profile represent a new starting point in the redesign and in-cell recruitment of molecules for therapeutic purposes [93–97]. Prior to the discovery of cell-penetrating peptides (CPPs), various methods have been used for cellular uptake of therapeutic agents and drugs, such as microinjection, electroporation, and liposome- and viral-based vectors, but these have disadvantages such as restricted bioavailability, low productivity, high toxicity, and low specificity [95]. After all these developments, in the late 1980s, a group of short peptides, such as the protein translocation site, membrane translocation sequence, Trojan peptide, or most commonly CPP, which serve as cellular uptake and delivery vectors of large molecules for therapeutic purposes, were identified [94, 96].

5.1 Definition and classification of CPP

CPPs are mostly defined as the short (containing less than 40 amino acid residues) partially hydrophobic and/or polybasic natural and synthetic peptides [94, 97]. With the discovery of CPPs, it has emerged as a new tool that allows cell membrane translocation without significant membrane damage and at low concentrations without using any chiral receptors. Due to the penetrating properties of these effective cell-penetrating peptides, it opened new opportunities for medical treatment and delivery of therapeutic agents across cells in vivo. Furthermore, and even more importantly, interdisciplinary studies show that CPPs are able to assist in internalizing covalently or electrostatically conjugated bioactive cargos such as nucleic acids, polymers, liposomes, nanoparticles, and low-molecular-weight drugs, with dose-dependent efficiency, cheapness, degradability inside cells, and low toxicity [94, 95, 97]. There are various classifications that rely on CPPs' qualities. A categorization based on the origin of the peptide is one of the classifications. The first subgroup is also called protein transduction domains (PTDs); they are including protein-derived peptides such as TAT and penetratin. The second subgroup peptides called chimeric peptides may occur two or more motifs from other peptides. And the last subgroup in this category is synthetic peptides such as the polyarginine family [94, 98]. In addition, CPPs are categorized according to peptide sequences and binding abilities to the lipids, including non-amphipathic (naCPPs), primary amphipathic (paCPPs), and secondary amphipathic (saCPPs) CPPs. R9 and TAT (48–60) are examples of naCPPs that have a high cationic amino acid ingredient, and they are bound to the cell membrane with a high consideration of anionic lipids. The second subgroup of paCCPs such as transportan or TP10 have hydrophobic and hydrophilic residues straight from their primary structure. The third subgroup in this categorization is saCPPs such as penetratin, pVEC, and M918. Their alpha-helix or beta-sheet patterns compose amphipathic properties [94].

In summary, in vitro cell and membrane modeling studies have shown that paCPPs are more toxic than naCPP and saCPPs.

5.2 Cellular uptake mechanism of CPP

5.2.1 Direct penetration

Direct penetration of CPPs eventuates along the membrane lipid bilayers without the contribution of receptors in low temperatures via energy-independent mechanism [93, 99]. There are three different types for the internalization pathway of CPPs by direct penetration [93].

5.2.1.1 Barrel-stave model

In the barrel-stave model, a different number of peptides are placed in the inside part of the membrane to give a hoop appearance. Each of these peptide molecules that provide the formation of membrane channels is likened to the boards that make up the barrel. Due to the appearance of the membrane, this model is called as the barrel-stave model. The barrel-stave model is carried out by an accumulation of peptides consisting of cationic peptides in the amphipathic helix structure. In this model, the hydrophilic sides of the peptides are contacted with the solvent, and the hydrophobic sides of the peptide come into contact with the lipid bilayer of the membrane, resulting in channels or pores in the membrane. The critical step in the barrel-stave model is that the peptides recognize each other while they are attached to the membrane. The peptide molecules then interact with the lipid molecules present in the membrane, and the peptides pass to the hydrophobic side of the membrane. By adding new monomers or small oligomers to the resulting structure, the width of the pores increases. The peptides act in this way even at very low concentrations once they penetrate the phospholipid membrane in the target cell [93].

5.2.1.2 Carpet-like model

According to the carpet-like model, firstly an electrostatic interaction occurs between the phospholipid groups on the negatively charged cell membrane bilayer and the positively charged peptide monomers. The hydrophilic sides of the peptide monomers are bound to the surface of the target cell in such a way that they overlap the phospholipid groups in the membrane and cover the cell as a carpet. Once the peptides in this state have reached a sufficient concentration, the peptide molecules rotate, reorganizing the hydrophobic portion of the membrane, causing a similar effect to the detergents and resulting in digestion of the membrane. On the contrary to the model of the barrel-stave model in the carpet model, the positively charged peptides cannot penetrate the hydrophobic parts of the membrane but instead are in a group with the hydrophilic parts facing one another. According to this mechanism, CPPs can be present in various secondary structures, sizes, and linear or cyclic form [93].

5.2.1.3 Toroidal model

In toroidal model, after the halezonal peptides are parallel to the membrane, the polar groups of the membrane are replaced by the hydrophobic amino acid groups of the peptides. This leads to the formation of cleft that is vertical to the membrane in the hydrophobic part of the membrane, extending along the peptide and lipid groups. The difference of the toroidal model and the barrel-stave model is that the peptides are present vertically in the lipid membrane and always form the pores in combination with the lipid groups [93, 100].

Direct penetration pathway results in an irreversible membrane destabilization and can cause artifact distribution in cells [93, 99]. Therefore, it has been shown that CPP-cargo complexes can be taken into the cell by a mechanism known as endocytosis [93, 101].

5.2.2 Endocytosis

Endocytosis is a natural process that occurs in all cells via energy-dependent mechanism. Different types of pinocytic pathways have been identified, since they vary according to the access of different CPPs and conjugated cargo complexes of CPPs [96].

5.2.2.1 Macropinocytosis

Macropinocytosis, which is related to the folding of the outer surface of the plasma membrane, is a nonspecific uptake of the extracellular compounds. It starts with an invasion of a membrane promoted by actinic cytoskeleton elements to create a pocket and large endocytic vesicles containing different types of cargoes, and it finishes by the occurrence of vesicles called macropinosomes. Macropinocytosis has critical importance in the uptake of TAT and polyarginine [93, 102]. TAT uptake happens when using macropinocaytosis without the need for GATs or sialic acids, while plasma membrane-active proteins must be present [103]. Passive transport does not have actin remodeling, so passive transport is recommended for CPP penetration instead of macropinocytosis [104].

5.2.2.2 Clathrin-mediated endocytosis

Clathrin-mediated endocytosis is the specific uptake of CPPs that have been shown to have significance in the uptake of penetratine, TAT peptide, and other

CPPs after inhibition by hyperosmolar media [96]. The mechanism starts with the creation of a vesicle that is covered by a crystal on the interior surface of the cell membrane. After the coating of clathrin proteins by vesicles, it combines with an early endosome which moves via microtubules from the cell into the nucleus. Macromolecules carried on late endosome are fused with vesicles of the Golgi apparatus, which include pioneer of lysosomal hydrolase. After combining, lysosomal hydrolase enzymes are activated, and the late endosome turns into active lysosome. In the lysosome, the endocytosis substance is decomposed [105].

5.2.2.3 Caveolin-mediated endocytosis

Caveolin-mediated endocytosis which starts with a flask-shaped pit in a cave-like membrane is a specific uptake of extracellular molecules. The CPP with its cargo is bound to the membrane, trapped in caveolae which are linked to actinic cytoskeletal elements, and then a protein tyrosine phosphorylation and actin depolymerization are supported. Dynamin is the type of a GTPase liable for endocytosis in the eukaryotic cells that define other actin polymerization on the patch. Finally, cargo-loaded pouches are released inside the cytosol [106]. The various endocytic paths mentioned above have disadvantages such as low specificity, high dependence on the cell line used, and decreased cell viability [96, 107].

5.2.2.4 Escape from endosomes

Endocytosis of CPPs happens in two stages as endocytic entry and endosomal escape. Endosomal escape is a major step to prevent disruption of the cargo in the lysosomes and to allow the cargo to reach the extra-endosomal aim and to achieve its biological activity. Several models have been proposed to prevent endosomal escapes of CPPs. For instance, as a result of the electrical interaction between the negative charges of the endosomal membrane and the positively charged parts of the CPPs, the membrane that triggers the release of the vesicle content leads to stiffing and rupture. According to the other example, reduced pH develops the capability of CPPs to attach to the endosomal membrane and then raises intracellular circulation; as a result of which, the increment of endocytic vesicles may ensure better escape from the endosomal membrane [93, 95].

5.3 Factors affecting the mechanism of cellular uptake

5.3.1 Role of glycosaminoglycans

Initially the CPPs and the cell membrane interact via electrostatic linkage to proteoglycans called as glycosaminoglycans (GAGs) such as chondroitin sulfate B. Since these glycoproteins are present all over the cell membrane, they make the cell membrane negative charges, and they compose a platform that connects the CPPs or CPP/cargo binding to the cellular matrix. Recent studies have shown that GAGs are less critical for CPP uptake than the previous opinions [93, 103, 108].

5.3.2 Peptide secondary structure

The secondary structure of amphipathic and tachycardic peptides, the CPP conformation, and the length of the CPP sequence are very related to cellular uptake mechanisms.

5.3.3 Role of arginine-rich residues

Structure-activity relationship studies show how important it is to identify the single residues in the CPP structure. When the CPPs are rich in arginine (especially for its guanidinium group), they can compose hydrogen bonds with polar lipid groups. The presence of arginines in the CPP has been related with a better uptake efficiency. In addition, replacement or wiping of arginines may reduce the cellular uptake [93, 104, 106].

5.3.4 Role of chirality

Although L amino acid and D amino acid peptides have similar effects against heparin, cell binding affinities of CPPs containing with amino acids in the L amino acid peptides are higher than D amino acid peptides. On the other hand, studies show that using of the D amino acid peptides in terms of increasing cell sensitivity and decreasing enzyme degradation gives better results [93, 101].

5.3.5 Role of concentration

It has been understood from the result of thermodynamic analyses that primary and secondary amphipathic CPPs can pass directly through the cell membrane at low micromolar concentrations, but non-amphipathic CPPs often use endocytosis even at low concentrations [109]. The concentration threshold of direct penetration depends on the type of CPPs, cargo, and cell lines [93].

5.3.6 Role of cargos

The existence of the cargo could affect the CPP uptake pathway. In addition, dimension and binding methodology are also shown by CPP translocation mechanism.

5.4 Molecular detection of CPPs' cellular uptake

5.4.1 In cellular assay

The cellular researches largely propose to follow the CPP and/or the cargo uptake or to elicit the molecular mechanisms of the internalization. As a direct method, matrix-assisted laser desorption/ionization time-of-flight mass spectrometry (MALDI-TOF-MS) has been improved, and quantitatively determination of the amount of intact CPPs in the cells or in the cellular membranes can be provided [110]. Electron microscopy and Raman spectroscopy methods are also other biophysical methods used on cells [111].

5.4.2 Fluorescence-based assays

The covalent binding of a peptide to the fluorophore and the measurement of the fluorescence (fluorimetry) of the treated cells are the most common methods used to detect both CPP uptake and localization. In this approach, while confocal microscopy provides location information, the probes in the living cells allow indirect quantification of the peptides. This method cannot show the molecular integrity of internalized entities, so it cannot prove whether the peptide is still attached to the fluorescent probe in the cells. Furthermore, fluorescence quenching can cause false-positive outcomes to detect the involvement in intracellular parts and transitory binding areas [94, 96, 111]. Also, CPPs are known as cationic peptides, and they tend to link to the external of the cell membrane and may consequently lead to false-positive results [96]. Fluorescence-activated cell sorting (FACS) is another fluorescence approach used for quantitative measurement uptake of labeled CPPs. The cellular uptake of CPPs is measured by sorting cells which relied on fluorescence intensity by a cell sorter equipment [94]. The advantage of using the aforementioned fluorescence-based protocols in the confocal microscopy process is that it allows distinguishing between internalized and extracellular peptides [94, 96].

5.4.3 Functional assays

These experiments, which are really suitable for biotechnological and therapeutic applications, can be carried out to determine the biological responses of linked molecules or cargos and to assess their mechanisms of uptake [1, 94]. These systems are based on internalizing antisense oligonucleotides (ONs) and upregulation of luciferase gene expression [96, 112]. One of the functional assays is the splicing redirection assay. Due to elucidation of the mechanism of CPP-based cell internalization, cell lines are transfected by molecular methods with a plasmid carrying a luciferase gene and having their abnormal splice junction. Technically, the pre-mRNA of luciferase is correctly working, producing the expression of the protein whose activity can be utilized by luminescence when the abnormal joint site is blocked by antisense ONs [94, 96, 112]. Another system takes advantage of Cre-mediated recombination of an improved green fluorescent protein reporter gene causing EGFP expression [94, 96]. Since abnormal splicing of the luciferase pre-mRNA inhibits the translation of luciferase, if this junction is masked with antisense ON, translation of luciferase will result in the expression of activity of the protein, which can be assessed by luminescence. For the investigation of CPP-based transfer of different ONs, their internalization systems have been studied with transfected HeLa pLuc 705 cells [96, 112].

5.4.4 MALDI-TOF-MS assay

This approach is based on the quantification of internalization of peptides in living cells [111]. Furthermore, the quantity of both plasma membrane-bound and internalized peptides is accessible. In this method, an internal standard is needed for the ionization and the detection in MALDI which extremely depends on the structure of the molecule. Information about both the amount and molecular status (degraded or intact) of the internalized peptide can be obtained by using this method. The disadvantage of this method is that no data about the intracellular localization of the peptide can be obtained. Nevertheless this method can get a knowledge about mechanisms of internalization (endocytosis versus direct translocation) and compare the relevant internalization affinities of CPP [96].

5.4.5 Membrane modeling assay

The assay often prefers to simulate the interaction, internalization, and direct translocation of CPPs through the inert plasma membrane [94, 96]. Membrane models are used in other techniques such as 31P-NMR, small-angle X-ray scattering, calcein or fluorescein-entrapped liposome leakage, and differential scanning calorimetry. These methods are not based on labeling of the CPP sequences with fluorescence agents [1]. Various researches have been performed by using giant or large unilamellar vesicles (GUV and LUV) to examine the translocation qualities of CPPs [96].

Unfortunately, the precise mechanism of CPP internalization is still not fully elucidated. CPPs are internalized into the cells by direct or endosomal mechanisms, and these uptake mechanisms have been demonstrated by various molecular techniques. Though many recent studies have provided strong data for translocation of CPPs into cells, a better physical definition of these entry mechanisms in cells by molecular techniques is still an open challenge for the future.

6. Conclusion

Consequently, this chapter provides a brief manual for anyone in the fields of solid-phase peptide synthesis, peptide vaccines, peptide theranostics, and cell-penetrating peptides. While the solid-phase peptide synthesis is the most convenient way to synthesize the peptides, the need for peptides with longer chain structure and peptides including unnatural amino acids has emerged with the identification of the pathological and physiological functions of the peptides through improvements in the molecular biology, bioengineering, and medical imaging areas. Synthetic chemistry will overcome these challenges, and peptides produced for different purposes, such as drugs, vaccines, and therapeutics, will be the solution for incurable diseases such as lots of epidemic disease and some cancer types with available materials and methods. It has been foreseen that the above-mentioned researches can be investigated within the peptide science and technology departments to be established academically in the next decade.

Conflict of interest

The authors declare no conflict of interest.

Author details

Burcu Ucar[1*], Tayfun Acar[1], Pelin Pelit Arayici[1], Melis Sen[2], Serap Derman[1] and Zeynep Mustafaeva[1]

1 Yildiz Technical University, Istanbul, Turkey

2 Koc University Hospital, Istanbul, Turkey

*Address all correspondence to: brccr87@gmail.com

References

[1] Fosgerau K, Hoffmann T. Peptide therapeutics: Current status and future directions. Drug Discovery Today. 2015;**20**(1):122-128. DOI: 10.1016/j.drudis.2014.10.003

[2] Craik DJ, Fairlie DP, Liras S, Price D. The future of peptide-based drugs. Chemical Biology & Drug Design. 2013;**81**(1):136-147. DOI: 10.1111/cbdd.12055

[3] Okarvi S. Peptide-based radiopharmaceuticals and cytotoxic conjugates: Potential tools against cancer. Cancer Treatment Reviews. 2008;**34**(1):13-26. DOI: 10.1016/j.ctrv.2007.07.017

[4] Mercer DK, O'neil DA. Peptides as the next generation of anti-infectives. Future Medicinal Chemistry. 2013;**5**(3):315-337. DOI: 10.4155/fmc.12.213

[5] Gori A, Longhi R, Peri C, Colombo G. Peptides for immunological purposes: Design, strategies and applications. Amino Acids. 2013;**45**(2):257-268. DOI: 10.1007/s00726-013-1526-9

[6] Trier NH, Hansen PR, Houen G. Production and characterization of peptide antibodies. Methods. 2012;**56**(2):136-144. DOI: 10.1016/j.ymeth.2011.12.001

[7] Robinson JA. Max Bergmann lecture protein epitope mimetics in the age of structural vaccinology. Journal of Peptide Science. 2013;**19**(3):127-140. DOI: 10.1002/psc.2482

[8] Carpino LA, Ghassemi S, Ionescu D, Ismail M, Sadat-Aalaee D, Truran GA, et al. Rapid, continuous solution-phase peptide synthesis: Application to peptides of pharmaceutical interest. Organic Process Research & Development. 2003;**7**(1):28-37. DOI: 10.1021/op0202179

[9] Mäde V, Els-Heindl S, Beck-Sickinger AG. Automated solid-phase peptide synthesis to obtain therapeutic peptides. Beilstein Journal of Organic Chemistry. 2014;**10**(1):1197-1212. DOI: 10.3762/bjoc.10.118

[10] Dunn BM. Peptide Chemistry and Drug Design. Hoboken, New Jersey: John Wiley & Sons; 2015

[11] Atherton E, Sheppard R. Solid Phase Peptide Synthesis: A Practical Approach. Oxford: IRL Press; 1989

[12] Behrendt R, White P, Offer J. Advances in Fmoc solid-phase peptide synthesis. Journal of Peptide Science. 2016;**22**(1):4-27. DOI: 10.1002/psc.2836

[13] Moss JA. Guide for resin and linker selection in solid-phase peptide synthesis. Current Protocols in Protein Science. 2005;**40**(1):18.7.1-18.7.9. DOI: 10.1002/0471140864.ps1807s40

[14] Labadie JW. Polymeric supports for solid phase synthesis. Current Opinion in Chemical Biology. 1998;**2**(3):346-352. DOI: 10.1016/S1367-5931(98)80008-2

[15] Merrifield B. The role of the support in solid phase peptide synthesis. British Polymer Journal. 1984;**16**(4):173-178. DOI: 10.1002/pi.4980160404

[16] Mitchell A, Erickson B, Ryabtsev M, Hodges R, Merrifield R. Tert-butoxycarbonylaminoacyl-4-(oxymethyl) phenylacetamidomethyl-resin, a more acid-resistant support for solid-phase peptide synthesis. Journal of the American Chemical Society. 1976;**98**(23):7357-7362. DOI: 10.1021/ja00439a041

[17] Matsueda GR, Stewart JM. A p-methylbenzhydrylamine resin for improved solid-phase synthesis of peptide amides. Peptides.

1981;**2**(1):45-50. DOI: 10.1016/S0196-9781(81)80010-1

[18] Bonkowski B, Wieczorek J, Patel M, Craig C, Gravelin A, Boncher T. Basic concepts of using solid phase synthesis to build small organic molecules using 2-chlorotrityl chloride resin. ChemInform. **2015;1**(4):1-4. DOI: 10.4172/2329-6798.1000113

[19] Shelton PT, Jensen KJ. Linkers, resins, and general procedures for solid-phase peptide synthesis. In: Peptide Synthesis and Applications. New York: Springer; 2013. pp. 23-41. DOI: 10.1007/978-1-62703-544-6_2

[20] Varkey JT, Pillai VR. Synthesis of thioredoxin partial sequences on 1, 6-hexanedioI diacrylate (HDODA)-cross-linked polystyrene resin. The Journal of Peptide Research. 1998;**51**(1):49-54

[21] Howl J. Peptide Synthesis and Applications. NJ, Totowa: Springer Science & Business Media; 2005

[22] Chan W, White P. Fmoc Solid Phase Peptide Synthesis: A Practical Approach. Oxford: OUP; 1999

[23] Bolt H, Cobb S. A practical method for the synthesis of peptoids containing both lysine-type and arginine-type monomers. Organic & Biomolecular Chemistry. 2016;**14**(4):1211-1215. DOI: 10.1039/C5OB02279G

[24] Isidro-Llobet A, Alvarez M, Albericio F. Amino acid-protecting groups. Chemical Reviews. 2009;**109**(6):2455-2504. DOI: 10.1021/cr800323s

[25] Behrendt R, Huber S, Martí R, White P. New t-butyl based aspartate protecting groups preventing aspartimide formation in fmoc SPPS. Journal of Peptide Science. 2015;**21**(8):680-687. DOI: 10.1002/psc.2790

[26] Veber D, Milkowski J, Varga S, Denkewalter R, Hirschmann R. Acetamidomethyl. A novel thiol protecting group for cysteine. Journal of the American Chemical Society. 1972;**94**(15):5456-5461. DOI: 10.1021/ja00770a600

[27] Royo M, Alsina J, Giralt E, Slomcyznska U, Albericio F. S-phenylacetamidomethyl (phacm): An orthogonal cysteine protecting group for Boc and Fmoc solid-phase peptide synthesis strategies. Journal of the Chemical Society, Perkin Transactions. 1995;**1**(9):1095-1102. DOI: 10.1039/P19950001095

[28] Dekan Z, Mobli M, Pennington MW, Fung E, Nemeth E, Alewood PF. Total synthesis of human hepcidin through regioselective disulfide-bond formation by using the safety-catch cysteine protecting group 4, 4′-dimethylsulfinylbenzhydryl. Angewandte Chemie. 2014;**126**(11): 2975-2978. DOI: 10.1002/ange.201310103

[29] Brailsford JA, Stockdill JL, Axelrod AJ, Peterson MT, Vadola PA, Johnston EV, et al. Total chemical synthesis of human thyroid-stimulating hormone (hTSH) β-subunit: Application of arginine-tagged acetamidomethyl (AcmR) protecting groups. Tetrahedron. 2018;**74**(15):1951-1956. DOI: 10.1016/j.tet.2018.02.067

[30] McCurdy S. The investigation of Fmoc-cysteine derivatives in solid phase peptide synthesis. Peptide Research. 1989;**2**(1):147-152. DOI: 10.1007/s00726-014-1696-0

[31] Stathopoulos P, Papas S, Sakka M, Tzakos AG, Tsikaris V. A rapid and efficient method for the synthesis of selectively S-trt or S-mmt protected cys-containing peptides. Amino Acids. 2014;**46**(5):1367-1376. DOI: 10.1007/s00726-014-1696-0

[32] Huang H, Rabenstein D. A cleavage cocktail for methionine-containing

peptides. The Journal of Peptide Research. 1999;**53**(5):548-553. DOI: 10.1034/j.1399-3011.1999.00059.x

[33] Sieber P. Modification of tryptophan residues during acidolysis of 4-methoxy-2,3,6-trimethylbenzenesulfonyl groups. Effects of scavengers. Tetrahedron Letters. 1987;**28**(15):1637-1640. DOI: 10.1016/S0040-4039(00)95379-6

[34] Dick F. Acid Cleavage/Deprotection in Fmoc/tBiu Solid-Phase Peptide Synthesis. Peptide Synthesis Protocols. Springer; 1994. pp. 63-72. DOI: 10.1385/0-89603-273-6:63

[35] Acar T, Arayıcı PP, Ucar B, Karahan M, Mustafaeva Z. Synthesis, characterization and lipophilicity study of brucella abortus' immunogenic peptide sequence that can Be used in the future vaccination studies. International Journal of Peptide Research and Therapeutics. 2018:1-8. DOI: 10.1007/s10989-018-9739-0

[36] Ucar B, Acar T, Pelit-Arayici P, Demirkol MO, Mustafaeva Z. A new radio-theranostic agent candidate: Synthesis and analysis of (ADH-1)c-EDTA conjugate. Fresenius Environmental Bulletin. 2018;**27**(7):4751-4758

[37] Chakraborty AB, Berger SJ. Optimization of reversed-phase peptide liquid chromatography ultraviolet mass spectrometry analyses using an automated blending methodology. Journal of Biomolecular Techniques: JBT. 2005;**16**(4):327

[38] Greenfield NJ. Methods to estimate the conformation of proteins and polypeptides from circular dichroism data. Analytical Biochemistry. 1996;**235**(1):1-10

[39] Gopal R, Park JS, Seo CH, Park Y. Applications of circular dichroism for structural analysis of gelatin and antimicrobial peptides. International Journal of Molecular Sciences. 2012;**13**(3):3229-3244

[40] Yadav DK, Yadav N, SMP K. Vaccines: Present Status and Applications. Animal Biotechnology. Amsterdam: Elsevier; 2014. pp. 491-508. DOI: 10.1016/B978-0-12-416002-6.00026-2

[41] Siagian RC, Osorio JE. Novel approaches to vaccine development in lower-middle income countries. International Journal of Health Governance. 2018;**23**(4):288-300. DOI: 10.1108/IJHG-03-2018-0011

[42] Plotkin SA. Vaccines: The fourth century. Clinical and Vaccine Immunology. 2009;**16**(12):1709-1719. DOI: 10.1128/CVI.00290-09

[43] Moisa A, Kolesanova E. Synthetic peptide vaccines. Biochemistry (Moscow) Supplement Series B: Biomedical Chemistry. 2010;**4**(4):321-332. DOI: 10.1134/S1990750810040025

[44] Zhang R, Ulery BD. Synthetic vaccine characterization and design. Journal of Bionanoscience. 2018;**12**(1): 1-11. DOI: 10.1166/jbns.2018.1498

[45] Knittelfelder R, Riemer AB, Jensen-Jarolim E. Mimotope vaccination—From allergy to cancer. Expert Opinion on Biological Therapy. 2009;**9**(4):493-506. DOI: 10.1517/14712590902870386

[46] Nevagi JR, Toth I, Skwarczynski M. Peptide-based vaccines. In: Koutsopoulos S, editor. Peptide Applications in Biomedicine, Biotechnology and Bioengineering. Australia: Woodhead Publishing; 2018. pp. 327-358. DOI: 10.1016/B978-0-08-100736-5.00012-0

[47] Yang H, Kim DS. Peptide immunotherapy in vaccine development: From epitope to adjuvant. In: Advances in Protein Chemistry and Structural Biology. Vol. 99. Amsterdam:

Elsevier; 2015. pp. 1-14. DOI: 10.1016/bs.apcsb.2015.03.001

[48] Synthetic peptides as vaccines. In: Burdon RH, Knippenberg PHv, editors. Synthetic Polypeptides as Antigens. Amsterdam: Elsevier; Vol. 19. 1988. pp. 177-191. DOI: 10.1016/S0075-7535(08)70011-1

[49] Li W, Joshi M, Singhania S, Ramsey K, Murthy A. Peptide vaccine: Progress and challenges. Vaccine. 2014;**2**(3):515-536. DOI: 10.3390/vaccines2030515

[50] Marintcheva B. Lightning Round: Harnessing the Power of Viruses. Amsterdam: Elsevier; 2018. DOI: 10.1016/B978-0-12-810514-6.00008-8

[51] Skwarczynski M, Toth I. Peptide-based synthetic vaccines. Chemical Science. 2016;**7**(2):842-854. DOI: 10.1039/C5SC03892H

[52] Slingluff Jr CL. The present and future of peptide vaccines for cancer: Single or multiple, long or short, alone or in combination? Cancer Journal (Sudbury, Mass). 2011;**17**(5):343. DOI: 10.1097%2FPPO.0b013e318233e5b2

[53] Wagachchi D, Tsai J-YC, Chalmers C, Blanchett S, Loh JM, Proft T. PilVax—A novel peptide delivery platform for the development of mucosal vaccines. Scientific Reports. 2018;**8**(1):2555. DOI: 10.1038/s41598-018-20863-7

[54] Vakili B, Eslami M, Hatam GR, Zare B, Erfani N, Nezafat N, et al. Immunoinformatics-aided design of a potential multi-epitope peptide vaccine against leishmania infantum. International Journal of Biological Macromolecules. 2018;**120**:1127-1139. DOI: 10.1016/j.ijbiomac.2018.08.125

[55] Dowling DJ. Recent advances in the discovery and delivery of TLR7/8 agonists as vaccine adjuvants. ImmunoHorizons. 2018;**2**(6):185-197. DOI: 10.4049/immunohorizons.1700063

[56] Seroski DT, Hudalla GA. Self-assembled peptide and protein nanofibers for biomedical applications. In: Biomedical Applications of Functionalized Nanomaterials. Amsterdam: Elsevier; 2018. pp. 569-598. DOI: 10.1016/B978-0-323-50878-0.00019-7

[57] Cerezo D, J Pena M, Mijares M, Martínez G, Blanca I, B De Sanctis J. Peptide vaccines for cancer therapy. Recent Patents on Inflammation & Allergy Drug Discovery. 2015;**9**(1):38-45. DOI: 10.2174/1872213X09666150131141953

[58] Azmi F, Ahmad Fuaad AAH, Skwarczynski M, Toth I. Recent progress in adjuvant discovery for peptide-based subunit vaccines. Human Vaccines & Immunotherapeutics. 2014;**10**(3):778-796. DOI: 10.4161/hv.27332

[59] Singh K, Kaur J, Varshney GC, Raje M, Suri CR. Synthesis and characterization of hapten–protein conjugates for antibody production against small molecules. Bioconjugate Chemistry. 2004;**15**(1):168, 10.1021/bc034158v-173

[60] Ghosh M, Solanki AK, Roy K, Dhoke RR, Roy S. Carrier protein influences immunodominance of a known epitope: Implication in peptide vaccine design. Vaccine. 2013;**31**(41):4682, 10.1016/j.vaccine.2013.06.110-4688

[61] Kızılbey K, Mansuroğlu B, Derman S, Mustafaeva Akdeste Z. An in vivo study: Adjuvant activity of poly-n-vinyl-2-pyrrolidone-co-acrylic acid on immune responses against Melanoma synthetic peptide. Bioengineered. 2018;**9**(1):134-143. DOI: 10.1080/21655979.2017.1373529

[62] Brito LA, O'hagan DT. Designing and building the next generation of

improved vaccine adjuvants. Journal of Controlled Release. 2014;**190**:563, 10.1016/j.jconrel.2014.06.027-579

[63] Mustafaev MI. Polyelectrolytes in immunology: Fundamentals and perspectives. Turkish Journal of Chemistry. 1996;**20**(2):126-138

[64] Martins C, Sousa F, Araújo F, Sarmento B. Functionalizing PLGA and PLGA derivatives for drug delivery and tissue regeneration applications. Advanced Healthcare Materials. 2018;**7**(1):1701035. DOI: 10.1002/adhm.201701035

[65] Vartak A, Sucheck S. Recent advances in subunit vaccine carriers. Vaccine. 2016;**4**(2):12. DOI: 10.3390/vaccines4020012

[66] Fujita Y, Taguchi H. Nanoparticle-based peptide vaccines. In: Micro and Nanotechnology in Vaccine Development. Amsterdam: Elsevier; 2017. pp. 149-170

[67] Derman S, Mustafaeva ZA, Abamor ES, Bagirova M, Allahverdiyev A. Preparation, characterization and immunological evaluation: Canine parvovirus synthetic peptide loaded PLGA nanoparticles. Journal of Biomedical Science. 2015;**22**(1):89. DOI: 10.1186/s12929-015-0195-2

[68] Smith JD, Morton LD, Ulery BD. Nanoparticles as synthetic vaccines. Current Opinion in Biotechnology. 2015;**34**:217-224. DOI: 10.1016/j.copbio.2015.03.014

[69] Gomes A, Mohsen M, Bachmann M. Harnessing nanoparticles for immunomodulation and vaccines. Vaccine. 2017;**5**(1):6. DOI: 10.3390/vaccines5010006

[70] de Jong M, Verwijnen SM, de Visser M, Kwekkeboom DJ, Valkema R, Krenning EP. Peptides for Radionuclide Therapy. Targeted Radionuclide Tumor Therapy. Springer; 2008. pp. 117-144. DOI: 10.1007/978-1-4020-8696-0_7

[71] Fernandes AR, Oliveira A, Pereira J, Coelho PS. Nuclear medicine and drug delivery. In: Advanced Technology for Delivering Therapeutics. London: InTech; 2017. DOI: 10.5772/65708

[72] Stigbrand T, Carlsson J, Adams GP. Targeted Radionuclide Tumor Therapy: Biological Aspects. Netherlands: Springer; 2008

[73] Reisine T, Bell GI. Molecular biology of somatostatin receptors. Endocrine Reviews. 1995;**16**(4):427-442. DOI: 10.1210/edrv-16-4-427

[74] Modlin I, Pavel M, Kidd M, Gustafsson B. Somatostatin analogues in the treatment of gastroenteropancreatic neuroendocrine (carcinoid) tumours. Alimentary Pharmacology & Therapeutics. 2010;**31**(2):169-188. DOI: 10.1111/j.1365-2036.2009.04174.x

[75] Bozkurt MF, Virgolini I, Balogova S, Beheshti M, Rubello D, Decristoforo C, et al. Guideline for PET/CT imaging of neuroendocrine neoplasms with 68Ga-DOTA-conjugated somatostatin receptor targeting peptides and 18F–DOPA. European Journal of Nuclear Medicine and Molecular Imaging. 2017;**44**(9):1588-1601. DOI: 10.1007/s00259-017-3728-y

[76] Ginj M, Zhang H, Waser B, Cescato R, Wild D, Wang X, et al. Radiolabeled somatostatin receptor antagonists are preferable to agonists for in vivo peptide receptor targeting of tumors. Proceedings of the National Academy of Sciences. 2006;**103**(44):16436-16441. DOI: 10.1073/pnas.0607761103

[77] Fani M, Peitl P, Velikyan I. Current status of radiopharmaceuticals for

the theranostics of neuroendocrine neoplasms. Pharmaceuticals. 2017;**10**(1):30. DOI: 10.3390/ph10010030

[78] Barrio M, Fendler WP, Czernin J, Herrmann K. Prostate specific membrane antigen (PSMA) ligands for diagnosis and therapy of prostate cancer. Expert Review of Molecular Diagnostics. 2016;**16**(11):1177-1188. DOI: 10.1080/14737159.2016.1243057

[79] Fendler WP, Eiber M, Beheshti M, Bomanji J, Ceci F, Cho S, et al. 68 Ga-PSMA PET/CT: Joint EANM and SNMMI procedure guideline for prostate cancer imaging: Version 1.0. European Journal of Nuclear Medicine and Molecular Imaging. 2017;**44**(6):1014-1024. DOI: 10.1007/s00259-017-3670-z

[80] Tulsyan S, Das CJ, Tripathi M, Seth A, Kumar R, Bal C. Comparison of 68Ga-PSMA PET/CT and multiparametric MRI for staging of high-risk prostate cancer68Ga-PSMA PET and MRI in prostate cancer. Nuclear Medicine Communications. 2017;**38**(12):1094-1102. DOI: 10.1097/MNM.0000000000000749

[81] Demirkol MO, Acar Ö, Uçar B, Ramazanoğlu SR, Sağlıcan Y, Esen T. Prostate-specific membrane antigen-based imaging in prostate cancer: Impact on clinical decision making process. The Prostate. 2015;**75**(7):748-757. DOI: 10.1002/pros.22956

[82] Demirkol MO, Kiremit MC, Acar O, Ucar B, Saglican Y. The utility of 68Ga-PSMA PET/CT in poorly differentiated metastatic prostate cancer. Clinical Nuclear Medicine. 2017;**42**(5):403-405. DOI: 10.1097/RLU.0000000000001617

[83] Haberkorn U. PSMA ligands for diagnosis and therapy of prostate cancer. Cancer Imaging. 2014;(9022):14(Suppl 1). DOI: 10.1186/1470-7330-14-S1-O10

[84] Zhang J, Mao F, Niu G, Peng L, Lang L, Li F, et al. 68Ga-BBN-RGD PET/CT for GRPR and integrin αvβ3 imaging in patients with breast cancer. Theranostics. 2018;**8**(4):1121. DOI: 10.7150/thno.22601

[85] Reubi J, Waser B, Schaer J-C, Laissue JA. Somatostatin receptor sst1–sst5 expression in normal and neoplastic human tissues using receptor autoradiography with subtype-selective ligands. European Journal of Nuclear Medicine. 2001;**28**(7):836-846. DOI: 10.1007/s002590100541

[86] Nock BA, Maina T, Béhé M, Nikolopoulou A, Gotthardt M, Schmitt JS, et al. CCK-2/gastrin receptor–targeted tumor imaging with 99mTc-labeled minigastrin analogs. Journal of Nuclear Medicine. 2005;**46**(10):1727-1736

[87] Wild D, Béhé M, Wicki A, Storch D, Waser B, Gotthardt M, et al. [Lys40 (Ahx-DTPA-111In) NH2] exendin-4, a very promising ligand for glucagon-like peptide-1 (GLP-1) receptor targeting.Journal of Nuclear Medicine. 2006;**47**(12):2025-2033. DOI: 10.2967/jnumed.115.167445

[88] Luo Y, Pan Q, Yao S, Yu M, Wu W, Xue H, et al. Glucagon-like peptide-1 receptor PET/CT with 68Ga-NOTA-exendin-4 for detecting localized insulinoma: A prospective cohort study. Journal of Nuclear Medicine: Official Publication. Society of Nuclear Medicine. 2016;**57**(5):715. DOI: 10.2967/jnumed.115.167445

[89] Wester H. Pharmaceutical radiochemistry (I). In: Wester HJ editor. Munich Molecular Imaging Handbook Series. Scintomics. 2010;1

[90] Pruszynski M, Kang CM, Koumarianou E, Vaidyanathan G,

Zalutsky MR. D-amino acid peptide residualizing agents for protein radioiodination: Effect of aspartate for glutamate substitution. Molecules. 2018;**23**(5):1223. DOI: 10.3390/molecules23051223

[91] Dollé F, Schubiger P, Lehmann L, Friebe M. PET Chemistry—The Driving Force in Molecular Imaging. PA: Schubiger; 2007

[92] Heidari P, Szretter A, Rushford LE, Stevens M, Collier L, Sore J, et al. Design, construction and testing of a low-cost automated 68Gallium-labeling synthesis unit for clinical use. American Journal of Nuclear Medicine and Molecular Imaging. 2016;**6**(3):176

[93] Borrelli A, Tornesello A, Tornesello M, Buonaguro F. Cell penetrating peptides as molecular carriers for anti-cancer agents. Molecules. 2018;**23**(2):295. DOI: 10.3390/molecules23020295

[94] Madani F, Lindberg S, Langel Ü, Futaki S, Gräslund A. Mechanisms of cellular uptake of cell-penetrating peptides. Journal of Biophysics. 2011;**2011**:414729. DOI: 10.1155/2011/414729

[95] Copolovici DM, Langel K, Eriste E, Langel U. Cell-penetrating peptides: Design, synthesis, and applications. ACS Nano. 2014;**8**(3):1972-1994. DOI: 10.1021/nn4057269

[96] Bechara C, Sagan S. Cell-penetrating peptides: 20 years later, where do we stand? FEBS Letters. 2013;**587**(12):1693-1702. DOI: 10.1016/j.febslet.2013.04.031

[97] Nore BF. A synthetic cell-penetrating peptide (CPP) with protamine conjugate utilized for gene delivery. Journal of Sulaimani Medical College. 2012;**2**(1):1-7

[98] Zorko M, Langel Ü. Cell-penetrating peptides: Mechanism and kinetics of cargo delivery. Advanced Drug Delivery Reviews. 2005;**57**(4):529-545. DOI: 10.1016/j.addr.2004.10.010

[99] Ye J, Liu E, Yu Z, Pei X, Chen S, Zhang P, et al. CPP-assisted intracellular drug delivery, what is next? International Journal of Molecular Sciences. 2016;**17**(11):1892. DOI: 10.3390/ijms17111892

[100] Meng S, Xu H, Wang F. Research advances of antimicrobial peptides and applications in food industry and agriculture. Current Protein and Peptide Science. 2010;**11**(4):264-273. DOI: 10.2174/138920310791233369

[101] Guidotti G, Brambilla L, Rossi D. Cell-penetrating peptides: From basic research to clinics. Trends in Pharmacological Sciences. 2017;**38**(4):406-424. DOI: 10.1016/j.tips.2017.01.003

[102] Wadia JS, Stan RV, Dowdy SF. Transducible TAT-HA fusogenic peptide enhances escape of TAT-fusion proteins after lipid raft macropinocytosis. Nature Medicine. 2004;**10**(3):310. DOI: 10.1038/nm996

[103] Gump JM, June RK, Dowdy SF. Revised role of glycosaminoglycans in TAT protein transduction domain-mediated cellular transduction. Journal of Biological Chemistry. 2010;**285**(2):1500-1507. DOI: 10.1074/jbc.M109.021964

[104] Nakase I, Niwa M, Takeuchi T, Sonomura K, Kawabata N, Koike Y, et al. Cellular uptake of arginine-rich peptides: Roles for macropinocytosis and actin rearrangement. Molecular Therapy. 2004;**10**(6):1011-1022. DOI: 10.1016/j.ymthe.2004.08.010

[105] Mousavi SA, Malerød L, Trond B, Kjeken R. Clathrin-dependent endocytosis. Biochemical Journal. 2004;**377**(1):1-16. DOI: 10.1042/bj20031000

[106] Maiolo JR, Ferrer M, Ottinger EA. Effects of cargo molecules on the cellular uptake of arginine-rich cell-penetrating peptides. Biochimica et Biophysica Acta (BBA)-Biomembranes. 2005;**1712**(2):161-172. DOI: 10.1016/j.bbamem.2005.04.010

[107] Vercauteren D, Vandenbroucke RE, Jones AT, Rejman J, Demeester J, De Smedt SC, et al. The use of inhibitors to study endocytic pathways of gene carriers: Optimization and pitfalls. Molecular Therapy. 2010;**18**(3):561-569. DOI: 10.1038/mt.2009.281

[108] Raucher D, Ryu JS. Cell-penetrating peptides: Strategies for anticancer treatment. Trends in Molecular Medicine. 2015;**21**(9):560-570. DOI: 10.1016/j.molmed.2015.06.005

[109] Alhakamy NA, Berkland CJ. Polyarginine molecular weight determines transfection efficiency of calcium condensed complexes. Molecular Pharmaceutics. 2013;**10**(5):1940-1948. DOI: 10.1021/mp3007117

[110] Burlina F, Sagan S, Bolbach G, Chassaing G. A direct approach to quantification of the cellular uptake of cell-penetrating peptides using MALDI-TOF mass spectrometry. Nature Protocols. 2006;**1**(1):200. DOI: 10.1038/nprot.2006.30

[111] Ziegler A, Seelig J. High affinity of the cell-penetrating peptide HIV-1 tat-PTD for DNA. Biochemistry. 2007;**46**(27):8138-8145. DOI: 10.1021/bi700416h

[112] Kang S-H, Cho M-J, Kole R. Up-regulation of luciferase gene expression with antisense oligonucleotides: Implications and applications in functional assay development. Biochemistry. 1998;**37**(18):6235-6239. DOI: 10.1021/bi980300h

Chapter 3

Isoacylpeptide Method for Long-Chain and Difficult Sequence-Containing Peptide Preparation

Yoshio Hamada

Abstract

Peptides or small-size proteins are important substances for medicines, diagnosis, and molecular biology research. In organic synthesis, the peptide bonds formation is performed in an organic solution (liquid-phase peptide synthesis, LPPS), or on a resin (solid-phase peptides synthesis, SPPS). LPPS can prepare a high volume of peptides, but it is generally required long processes and high cost for peptides preparation and is not appropriate for long-chain peptides. SPPS can prepare long-chain peptides until 40 residues in a short time. However, it is difficult to obtain the pure peptides because of no purification of its intermediates. For a solution of these problems, Kent et al. reported native chemical ligation (NCL) method for the preparation of long-chain peptides. Because peptides with a long chain or difficult sequence formed β-sheet structure within a molecule, these peptides have high aggregability and low solubility, and their preparation and purification are generally difficult. Mutter et al. reported 'pseudoproline' method for difficult sequence-containing peptide preparation. We previously reported a series of prodrugs based on *O-N* intramolecular acyl migration. We reported '*O*-isoacylpeptide' method for the preparation of difficult sequence-containing peptides using the prodrug strategy based on *O-N* intramolecular acyl migration.

Keywords: difficult sequence-containing peptide, long-chain peptide, *O-N* intramolecular acyl migration, peptide synthesis, isoacylpeptide

1. Introduction

Peptides or small-size proteins are important substances for medicines, diagnosis, and molecular biology research, such as enzyme inhibitors, antagonists/agonists against receptors, antigenic peptides for antibody preparation, and peptide probes that detect a protein-peptide interaction. There are two general methodologies for peptide preparation—organic synthesis and genetic engineered synthesis. In organic synthesis, the peptide bonds formation is performed in an organic solution (liquid-phase peptide synthesis, LPPS), or on a resin (solid-phase peptides synthesis, SPPS). LPPS can prepare a high volume of peptides, but it is generally required long processes and high cost containing labor cost and is not appropriate for long-chain peptides. SPPS can prepare long-chain peptides until 40 residues in a short time.

 IntechOpen

However, it is difficult to obtain the pure peptides because of no purification of their intermediates. For a solution of these problem, Kent et al. reported native chemical ligation (NCL) method for the preparation of long-chain peptides as shown in **Figure 1A** [1–3]. In NCL reaction, a peptide possessing thioester at the C-terminus and a peptide possessing Cys residue at the N-terminus are prepared by SPPS. Next, both peptides are condensed by nucleophilic attack of thiol group at the N-terminus Cys residue in an aqueous solution, and then the condensed peptide with a thioester bond is spontaneously transformed into the peptide in which both peptides are connected with an amide bond. Although NCL allowed to preparing the long-chain peptides, this method is only available for the preparation of peptides with one or more Cys residues. Yan and Dawson [4] reported a modified NCL method for preparation of the peptides with one or more Ala residues (**Figure 1B**, $R^1 = R^2 =$ **—**H). In this reaction, the Cys-containing peptides that were obtained by NCL reaction are reduced into the Ala-containing peptides using Raney-Nickel catalyst. Haase et al. [5]

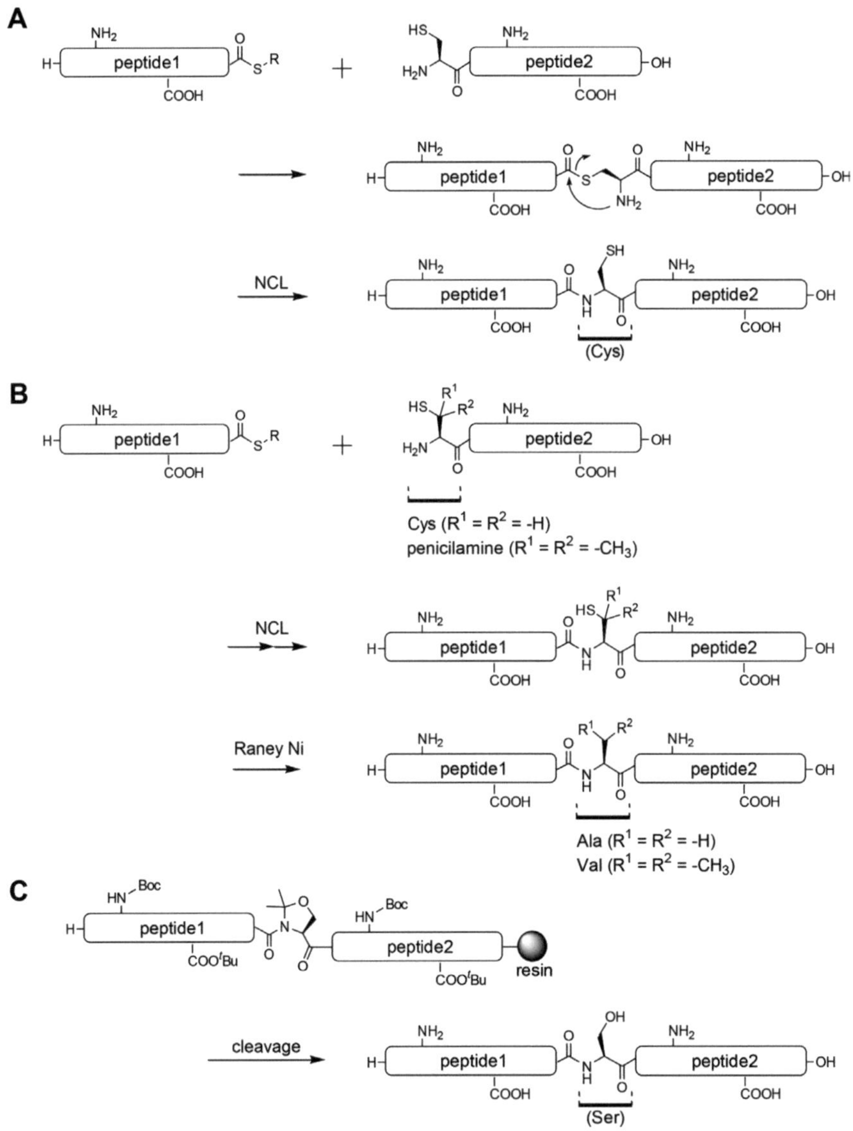

Figure 1.
(A) Native chemical ligation. (B) Expanded native chemical ligation. (C) Pseudoproline method.

reported other modified NCL method for preparation of the Val-containing peptides from peptides with one or more penicillamine (β-mercaptovaline) as shown in **Figure 1B** ($R^1 = R^2 = —CH_3$) in a manner similar to Haase et al.

There are synthetic difficult peptides containing specific amino acid sequences. Because the difficult sequence-containing peptides and long-chain peptides formed β-sheet structure within a molecule, these peptides have high aggregability and low solubility in aqueous and organic solvents, and their preparation and purification are generally difficult. Although we can use various resin for preparation of difficult sequence-containing peptides, such as Tentagel™ (RAPP Polymere, Germany) with a PEG moiety on the polystyrene (PS) bead and HDODA resin with a flexible crosslinker [6], some game-changing technologies for the preparation of difficult sequence-containing peptides has been reported. Mutter et al. reported 'pseudoproline' method for the preparation of difficult sequence-containing peptides and long-chain peptides [7, 8] as shown in **Figure 1C**. Mutter et al. synthesized peptides on a resin using an oxazolidine-containing amino acid (pseudoproline) in which α-amino and hydroxyl groups of Ser or Thr are cyclized by acetonidation. The deprotection and cleavage of peptides on the resin by a strong acid can convert form peptides with cyclic amino acid residue into peptides with Ser or Thr residue. Because the cyclic amino acids have a structure similar to Pro and peptides with a cyclic amino acid residue have greatly different structure compared with original peptides, peptides containing a pseudoproline are prevented to form β-sheet structure within a molecule. Because the oxazolidine-containing amino acids are labile in acid media, the dipeptide units, which consist of Fmoc-protected amino acid and oxazolidine-containing amino acid are commercially available.

We previously reported a series of prodrugs based on *O-N* intramolecular acyl migration [9–13]. Furthermore, we reported '*O*-isoacylpeptide' method for the preparation of difficult sequence-containing peptides using the prodrug strategy based on *O-N* intramolecular acyl migration [14, 15].

2. *O-N* intramolecular acyl migration and prodrugs

Previously, we reported a series of water-soluble prodrugs, such as human immunodeficiency virus type-1 (HIV-1) protease inhibitors and anti-cancer drugs [10–12, 15]. HIV-1 is a retrovirus that causes the acquired immunodeficiency syndrome (AIDS). Since HIV-1 encodes HIV-1 protease that is responsible for the processing of viral precursor proteins such as gag and gag-pol polyproteins to form mature structural proteins and some enzymes required in the production of infective viral particles, HIV-1 protease is an attractive target for the design of anti-AIDS drugs. HIV-1 protease is an aspartic protease that consists of a C2-symmetric homodimer, and its active site has some hydrophobic pockets. Thus most of the HIV-1 protease inhibitors that are optimized for the active site have high hydrophobicity and are sparingly water soluble. Many HIV-1 protease inhibitor formulations contain some solubilizers such as polyethylene glycol derivatives that often lead to unwanted side effects in clinical use. A water-soluble prodrug of amprenavir, fosamprenavir [16, 17], was approved by the US Federal Drug Administration (FDA) in 2003, and amprenavir was discontinued by the manufacture in 2004. Our previously reported HIV-1 protease inhibitors [18–21] also showed poorly water-solubility similar to amprenavir. Hence, we designed a series of water-soluble prodrug of HIV-1 protease inhibitor using a novel prodrug strategy. An acyl migration on the β-hydroxy-α-amino acid residue such as Ser and Thr in strong acids was well-known in peptide chemistry [22]. An *N*-acyl β-hydroxy-α-amino acid residue isomerizes into the *O*-acyl form in strong acids, and *O*-isoacylpeptide can regenerate the original peptide in weak alkaline media *via O-N* intramolecular acyl migration (**Figure 2A**). Since our HIV-1 protease inhibitors contain a β-hydroxy-α-amino acid residue, (2S,3S)-3-amino-2-hydroxy-4-phenylbutyric acid (Apns), we designed and synthesized the *O*-acyl isomer of KNI-727 as a water-soluble prodrug. The prodrug of KNI-727 was stable as an HCl salt in unbuffered aqueous solutions and in strong acidic solution such as gastric juice, and could be rapidly converted to the parent compound, KNI-727, *via O-N* intramolecular acyl migration reaction under the physiological condition

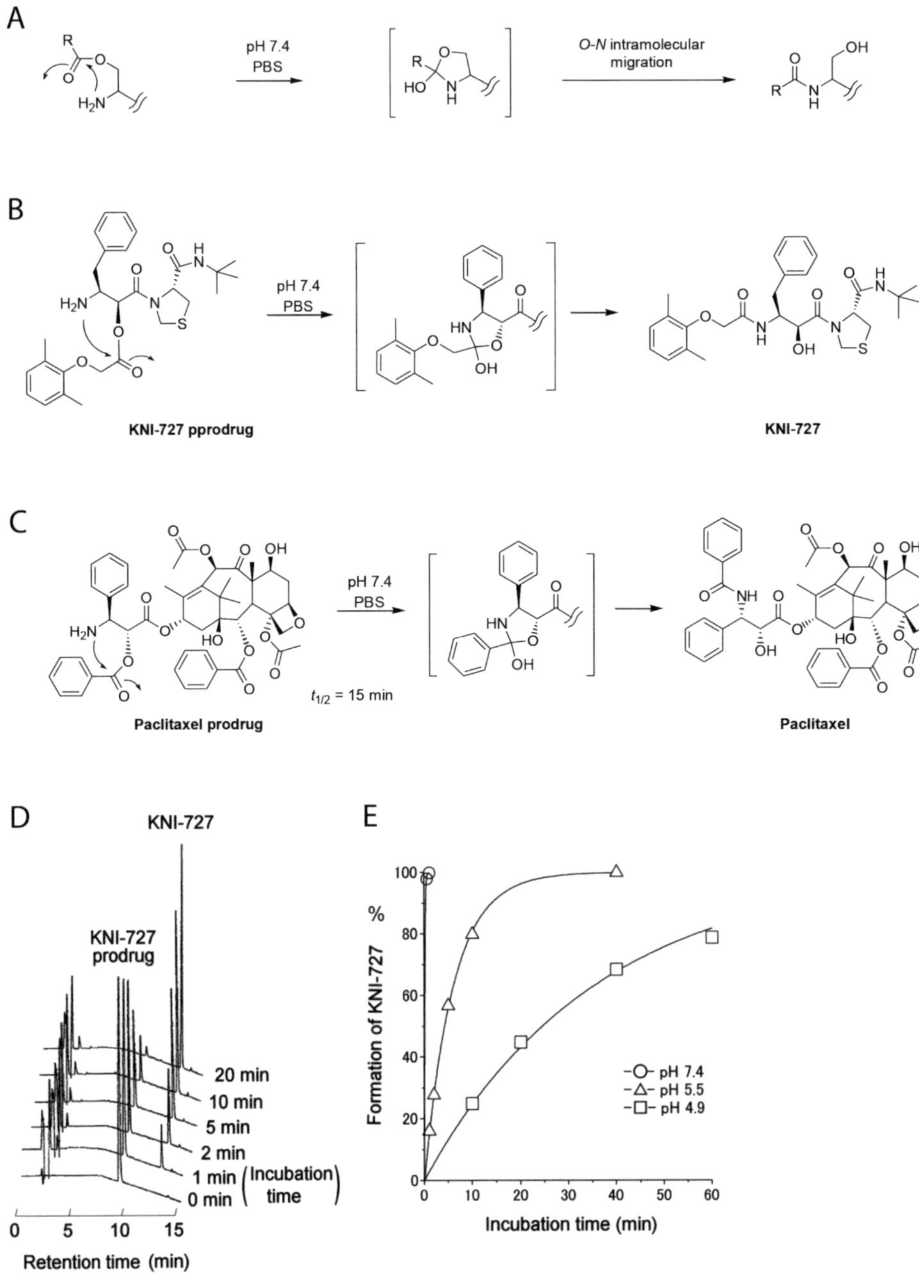

Figure 2.
Prodrugs based on O-N intramolecular acyl migration (A) O-N intramolecular acyl migration. (B) Prodrug of HIV-1 protease inhibitor, KNI-727. (C) Paclitaxel prodrug. (D) HPLC profile of the O-N intramolecular acyl migration of KNI-727 prodrug in PBS (pH 5.5) at 37°C. (E) Time course of the migration reaction of KNI-727 prodrug.

(pH 7.4 PBS, 37°C) as shown in **Figure 1B**. KNI-727 could be purified by preparative HPLC using a reverse-phase C18 column and a linear gradient system of acetonitrile and 0.1% TFA. HPLC charts and Time course of the migration reaction in KNI-727 prodrug were shown in **Figure 2D** and **E** respectively. The rate constant and $t_{1/2}$ were calculated using fitting Eq. (1).

$$[A]_t = A_{\mathrm{MAX}} \times (1 - \mathrm{Exp}(-k \times t))(1) \quad (1)$$

[t, incubation time; k, rate constant of migration; A_{MAX}, maximum concentration of the parent compound (initial concentration of prodrug); [A], concentration of the parent compound].

O-N intramolecular acyl migration of KNI-727 prodrug did not involve any by-product as shown in **Figure 2D** and **E** using the fitting Eq. (1) showed typical first order reaction. Hence, this water-soluble prodrug is suitable as an administrated drug for the AIDS therapy.

Next, we designed and synthesized water-soluble paclitaxel prodrug [12, 13]. Paclitaxel is an anti-cancer agent that was extracted from the Pacific yew tree *Taxus brevifolia* with antineoplastic activity. Since paclitaxel binds to tubulin, it can inhibit the disassembly of microtubules, thereby resulting in the inhibition of cell division, and also induces apoptosis by binding to the apoptosis inhibitor protein, B-cell leukemia 2 (Bcl-2). As paclitaxel are labile in acidic media such as gastric fluid, the paclitaxel formulations had been used as an injectable drug. The paclitaxel is representative of poorly water-soluble drug, and its injectable formulations require some detergents, such as Cremophor EL, which has been suggested to cause hypersensitivity. By focusing on the β-hydroxy-α-amino acid moiety in the chemical structure of paclitaxel, we designed the prodrug, *O*-benzoyl isopaclitaxel, in which the benzoyl group on the amino group of the paclitaxel was moved to its hydroxyl group. This paclitaxel prodrug is stable in water as a salt, and can be rapidly converted to the parent drug under physiological conditions ($t_{1/2}$ = 15 min, pH 7.4 PBS, 37°C) as shown in **Figure 2C**. The paclitaxel prodrug seems to be suitable as an injectable drug.

3. *O*-isoacylpeptide method

Since two natural amino acids, Ser and Thr, have a β-hydroxyl-α-amino acid structure, we designed the precursors of biomolecules, so-called '*O*-isoacylpeptides', using our prodrug strategy *via O-N* intramolecular acyl migration [14–16]. *O*-isoacylpeptides that have an *O*-acyl ester bond instead of amide bond in the Ser/Thr residue of biomolecules are promptly converted to the corresponding biomolecules under physiological conditions. It is known that some biomolecules are aggregate in aqueous solutions because of their β-sheet structure formation. Among them, Aβs show strong water-insolubility and aggregability, making their handling in biochemical research potentially complicated. Hence, chemical synthesis on resins and purification of Aβs, especially $A\beta_{1\text{-}42}$, is particularly challenging. To solve these problems, we designed *O*-isoacyl-$A\beta_{1\text{-}42}$ that has an ester bond instead of the amide bond at Gly_{25}-Ser_{26} in $A\beta_{1\text{-}42}$. Synthesis of $A\beta_{1\text{-}42}$ using *O*-isoacylpeptide method showed in **Figure 3A**. Peptide bonds formation on the PS-resin was performed by 9-fluorenylmethoxycarbonyl (Fmoc)-based SPPS using diisopropylcarbodiimide (DIPCDI) as a coupling reagent in the present of 1-hydroxybenzotriazole (HOBt). The hydroxy group in N-terminal Ser residue of the protected $A\beta_{27\text{-}42}$ that was prepared on a resin was esterified by Fmoc-Gly-OH using DIPCDI in the presence of catalytic amount of *N,N*-dimethyl-4-aminopyridine (DMAP). Next, remaining peptide bonds formation using the Fmoc-based SPPS similar to the former

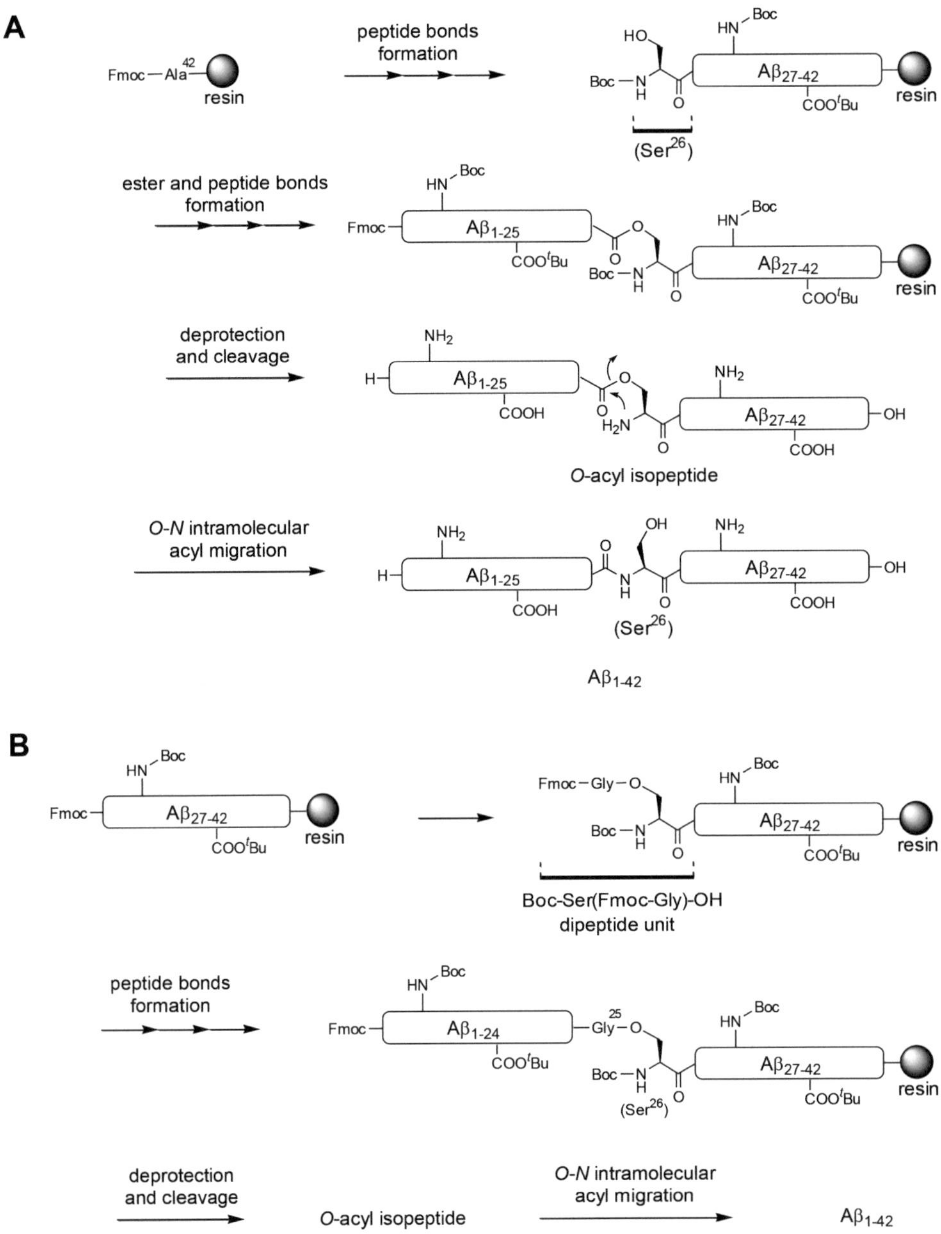

Figure 3.
O-isoacylpeptide method. (A) $A\beta_{1\text{-}42}$ synthesis via O-N intramolecular acyl migration. (B) $A\beta_{1\text{-}42}$ synthesis using an dipeptide unit.

procedure formed the protected *O*-isoacyl-$A\beta_{1\text{-}42}$, and then *O*-isoacyl-$A\beta_{1\text{-}42}$ was obtained by deprotection and cleavage from a resin. *O*-isoacyl-$A\beta_{1\text{-}42}$ was stable in acidic aqueous solution and unbuffered aqueous solution, and could rapidly release the native $A\beta_{1\text{-}42}$ peptide. Because *O*-isoacylisopeptides showed good stability in acidic media, *O*-isoacylisopeptides can be easily isolated and purification in acidic solution [23]. In this manner, *O*-isoacylpeptide method has the advantage over the 'pseudoproline' method. *O*-isoacyl-$A\beta_{1\text{-}42}$ could be easily synthesized on a resin and purified by preparative HPLC using a reverse phase C18 column in acidic eluent solvents, and could release native $A\beta_{1\text{-}42}$ that consists of $A\beta_{1\text{-}42}$ monomers. However, because ester bond formation on a resin often involves a racemization of protected amino acid, *O*-isoacyl-dipeptide units such as Boc-Ser(Fmoc-Gly)-OH have been commercial available from some chemical suppliers. *O*-isoacylpeptide method using an *O*-isoacyl-dipeptide unit was shown in **Figure 3B**. Use of Fmoc-*O*-isoacyl-dipeptide allows to synthesize the isoacylpeptides by the conventional Fmoc-based SPPS without use of DMAP catalyst.

Furthermore, we designed and synthesized photo cleavable-protected *O*-isoacylpeptide, so called 'photo-click peptide'. Because *O*-isoacylpeptides are rapidly converted to the biomolecules by stimuli such as pH changes and photo irradiation, as with the click of a button, we term 'click peptide'. Click peptides that can

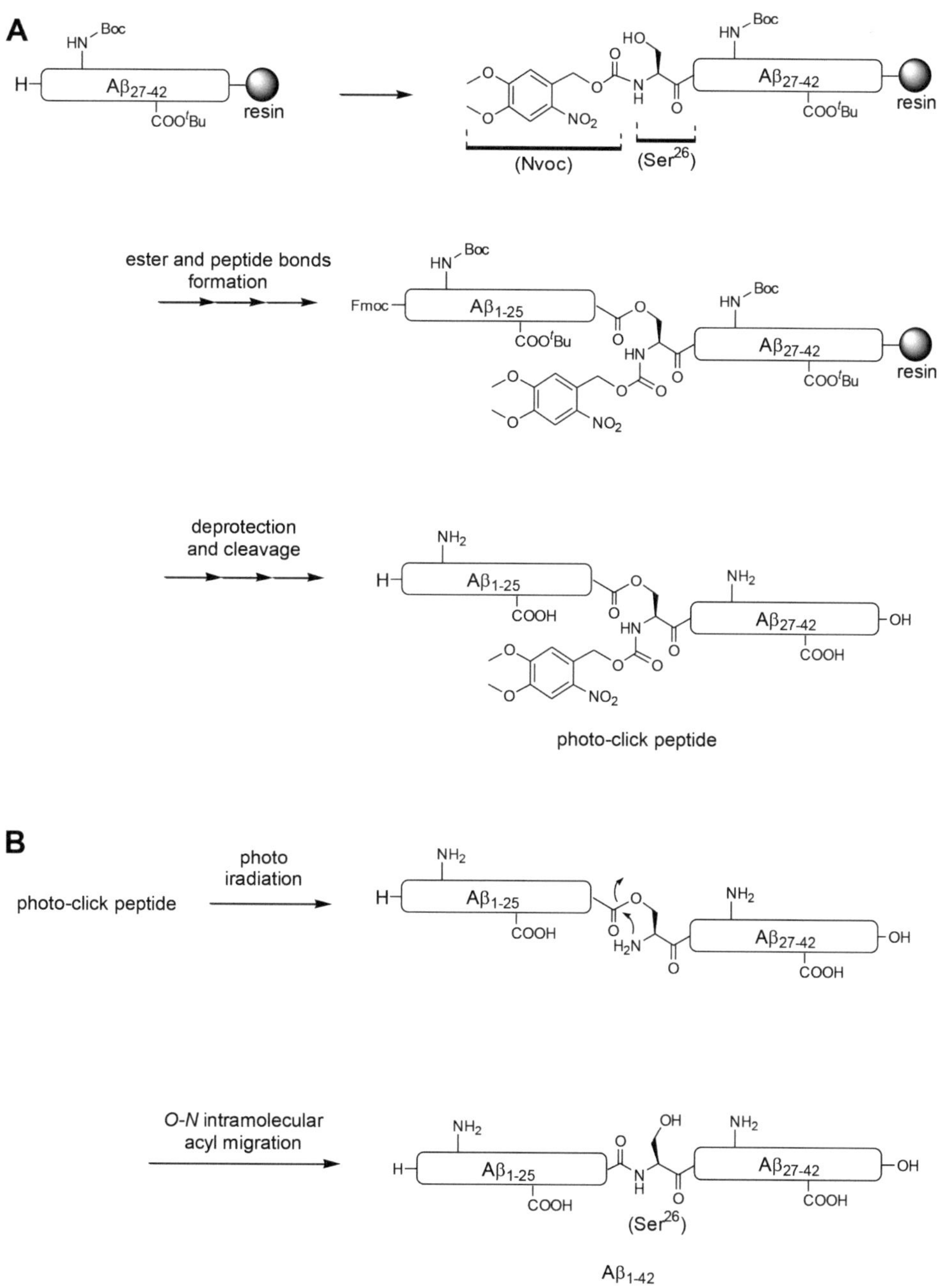

Figure 4.
Photo-click peptide. (A) Synthesis of photo-click $A\beta_{1-42}$. (B) Photo-click reaction.

generate bioactive molecules in situ *via* a 'click' appear to be useful tools for chemical biology research. Synthesis of photo-click peptide was shown in **Figure 4A**. Photo-click A$\beta_{1\text{-}42}$ could prepare on a resin in a similar manner showed in **Figure 3A**, using photo-cleavable protected amino acid instead of Boc-Ser-OH. Photo-cleavable protected amino acid, 6-nitroveratryloxycalbonyl (Nvoc)-Ser-OH, was coupled to protected A$\beta_{27\text{-}42}$ on the resin after Fmoc-deprotection, and then Fmoc-Gly-OH was coupled on the hydroxyl group of Ser$_{26}$ using DIPCDI in the presence of catalytic amount of DMAP. The remaining peptide bonds formation using the Fmoc-based SPPS formed the protected *O*-isoacyl-A$\beta_{1\text{-}42}$. Deprotection and cleavage from resin released *O*-isoacyl-A$\beta_{1\text{-}42}$. Photo-click A$\beta_{1\text{-}42}$ could rapidly release the native A$\beta_{1\text{-}42}$ peptide by photo-irradiation and subsequent *O-N* intramolecular acyl migration under physiological condition (**Figure 4B**).

4. Segment condensation of peptides using *O*-isoacylpeptides

Although NCL by Kent et al. allowed to preparing the long chain peptides, in general, conventional segment condensation other than NCL often involves the

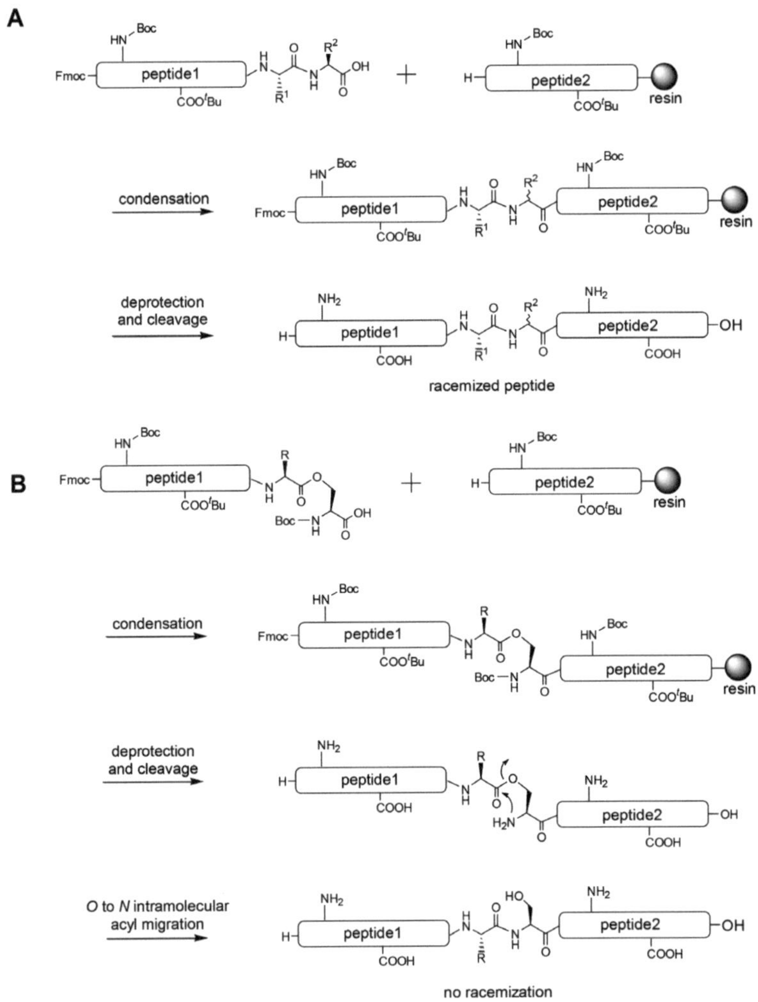

Figure 5.
Segment condensation of peptides. (A) Conventional segment condensation between peptides. (B) Segment condensation using O-isopeptide method.

racemization of amino acid as shown in **Figure 5A**. Especially, it is a serious problem in SPPS because of no purification of intermediates. It is well-known that the urethane structure, such as Boc and Fmoc protecting group, on the α-amino group of amino acid can prominently reduce the racemization of amino acid in peptide bond coupling reaction. Hence, segment condensation between peptides other than NCL must not be in peptide chemistry. We noticed the protected *O*-isoacylpeptide structure that possesses a urethane structure on the α-amino group of Ser or Thr residue. When an *O*-isoacylpeptide possessing a urethane-type protecting group at the C-terminus was coupled with another peptide, we speculated that the condensed peptide may be not racemized at the C-terminal amino acid of *O*-isoacylpeptide. Hence we designed segment condensation method as shown in **Figure 5B**. The segment condensation based on the *O*-isoacylpeptide method showed no racemization [24], and could release the original peptide similar to the other *O*-isoacylpeptides under physiological condition. This method appears to be alternative choice to NCL method.

5. Conclusion

Recently, some important synthesis methods such as NCL and pseudoproline method for preparation of long chain and difficult sequence-containing peptides had been reported. Although these approaches allow to preparing some long chain peptides without a genetic engineered approach. However, these methodologies are not a panacea for a long chain and difficult sequence-containing peptides. We supply alternative solution for the long chain and difficult sequence-containing peptide preparation. Namely, we have developed the *O*-isoacylpeptide method for peptide preparation. *O*-isoacylpeptides that have a/some ester bonds can be converted into the parent peptides under physiological condition *via* the *O-N* intramolecular acyl migration. Moreover, we developed segment condensation with no racemization using the *O*-isoacylpeptide method. The segment condensation based on *O*-isoacylpeptides may be alternative choice to NCL method. We desire to apply to these methodologies of various peptide preparations for drug, diagnosis, and molecular research.

Acknowledgements

This study was supported in part by the Grants-in-Aid for Scientific Research from MEXT (Ministry of Education, Culture, Sports, Science and Technology), Japan (KAKENHI No. 23590137, No. 26460163 and No.18K06562), and a donation from Mrs. Kazuko Fujita with the cherished desire of the late Tetsuro Fujita, Professor Emeritus of Kyoto University.

Conflict of interest

We confirmed independence from the funding source.

Author details

Yoshio Hamada
Faculty of Frontier of Innovative Research in Science and Technology, Konan University, Kobe, Japan

*Address all correspondence to: pynden@gmail.com

References

[1] Dawson PE, Muir TW, Clark-Lewis I, et al. Synthesis of proteins by native chemical ligation. Science. 1994;**266**:776-779. DOI: 10.1126/science.7973629

[2] Johnson ECB, Kent SBH. Insights into the mechanism and catalysis of the native chemical ligation reaction. Journal of the American Chemical Society. 2006;**128**:6640-6646. DOI: 10.1021/ja058344i

[3] Kent SBH. Total chemical synthesis of proteins. Chemical Society Reviews. 2009;**38**:338-351. DOI: 10.1039/b700141j

[4] Yan LZ, Dawson PE. Synthesis of peptides and proteins without cysteine residues by native chemical ligation combined with desulfurization. Journal of the American Chemical Society. 2001;**123**:526-533. DOI: 10.1021/ja003265m

[5] Haase C, Rohde H, Seitz O. Native chemical ligation at valine. Angewandte Chemie International Edition in English. 2008;**47**:6807-6810. DOI: 10.1002/anie.200801590

[6] Varkey JT, Pillai VNR. Merrifield resin and 1,6-hexanediol diacrylate-crosslinked polystyrene resin for solid-phase peptide synthesis: A comparative study. Journal of Applied Polymer Science. 1999;**71**:1933-1939. DOI: 10.1002/(SICI)1097-4628(19990321)71:12<1933::AID-APP1>3.0.CO;2-C

[7] Mutter M, Nefzi A, Sato T, et al. Pseudo-prolines (psi Pro) for accessing “inaccessible” peptides. Peptide Research. 1995;**8**:145-153

[8] Wöhr T, Wahl F, Nefzi A, et al. Pseudo-prolines as a solubilizing, structure-disrupting protection technique in peptide synthesis. Journal of the American Chemical Society. 1996;**118**:9218-9227. DOI: 10.1021/ja961509q

[9] Hamada Y, Ohtake J, Sohma Y, et al. New water-soluble prodrugs of HIV protease inhibitors based on O→N intramolecular acyl migration. Bioorganic & Medicinal Chemistry. 2002;**10**:4155-4167. DOI: 10.1016/S0968-0896(02)00322-X

[10] Hamada Y, Matsumoto H, Kimura T, et al. Effect of the acyl groups on O→N acyl migration in the water-soluble prodrugs of HIV-1 protease inhibitor. Bioorganic & Medicinal Chemistry Letters. 2003;**13**:2727-2730. DOI: 10.1016/S0960-894X(03)00576-6

[11] Hamada Y, Matsumoto H, Yamaguchi S, et al. Water-soluble prodrugs of dipeptide HIV protease inhibitors based on O→N intramolecular acyl migration: Design, synthesis and kinetic study. Bioorganic & Medicinal Chemistry. 2004;**12**:159-170. DOI: 10.1016/j.bmc.2003.10.026

[12] Hayashi Y, Skwarczynski M, Hamada Y, et al. A novel approach of water-soluble paclitaxel prodrug with no auxiliary and no byproduct: Design and synthesis of isotaxel. Journal of Medicinal Chemistry. 2003;**46**:3782. DOI: 10.1021/jm034112n

[13] Skwarczynski M, Sohma Y, Noguchi M, et al. No auxiliary, no byproduct strategy for water-soluble prodrugs of taxoids: Scope and limitation of O–N intramolecular acyl and acyloxy migration reactions. Journal of Medicinal Chemistry. 2005;**48**:2655-2666. DOI: 10.1021/jm049344g

[14] Sohma Y, Hayashi Y, Skwarczynski M, et al. O–N intramolecular acyl migration reaction in the development of prodrugs and the synthesis of difficult sequence-containing bioactive peptides. Biopolymers. 2004;**76**:344-356. DOI: 10.1002/bip.20136

[15] Hamada Y. Recent progress in prodrug design strategies based on generally applicable modifications. Bioorganic & Medicinal Chemistry Letters. 2017;**27**:1627-1632. DOI: 10.1016/j.bmcl.2017.02.075

[16] Hamada Y, Kiso Y. Aspartic protease inhibitors as drug candidates for treating various difficult-to-treat diseases. In: Amino Acids, Peptides and Proteins. Vol. 39. London: Royal Society of Chemistry; 2015. pp. 114-147

[17] Strickley RG. Solubilizing excipients in oral and injectable formulations. Pharmaceutical Research. 2004;**21**:201-230. DOI: 10.1023/B:PHAM.0000016235.32639.23

[18] Panov I, Drabina P, Hanusek J, et al. Stereoselective synthesis of the key intermediates of the HIV protease inhibitor fosamprenavir and its diastereomer. Synlett. 2013;**24**:1280-1282. DOI: 10.1055/s-0033-1338803

[19] Nguyen J-T, Hamada Y, Kimura T, et al. Design of potent aspartic protease inhibitors to treat various diseases. Archiv der Pharmazie—Chemistry in Life Sciences. 2008;**341**:523-535. DOI: 10.1002/ardp.200700267

[20] Hamada Y, Kiso Y. The application of bioisosteres in drug design for novel drug discovery: Focusing on acid protease inhibitors. Expert Opinion on Drug Discovery. 2012;7:903-922. DOI: 10.1517/17460441.2012.712513

[21] Hamada Y, Kiso Y. New directions for protease inhibitors directed drug discovery. Biopolymers. 2016;**106**: 563-579. DOI: 10.1002/bip.22780

[22] Hurley TR, Colson CE, Hicks G, et al. Orally active water-soluble N,O-acyl transfer products of α,β,γ-bishydroxyl amide containing renin inhibitor. Journal of Medicinal Chemistry. 1993;**36**:1496-1498. DOI: 10.1021/jm00062a024

[23] Taniguchi A, Sohma Y, Kimura M, et al. "Click peptide" based on the "O-acyl isopeptide method": Control of $A\beta_{1-42}$ production from a photo-triggered $A\beta_{1-42}$ analogue. Journal of the American Chemical Society. 2006;**128**:696-697. DOI: 10.1021/ja057100v

[24] Yoshiya T, Sohma Y, Hamada Y, et al. Racemization-free segment condensation based on the O-acyl isopeptide method: Toward a chemical protein synthesis on solid support. Advances in Experimental Medicine and Biology. 2009;**611**:161-162. DOI: 10.1007/978-0-387-73657-0_73

Chapter 4

Defensin-Like Peptides and Their Antimicrobial Activity in Free-Form and Immobilized on Material Surfaces

Marc Bruggeman, Hanieh Ijakipour and Artemis Stamboulis

Abstract

Defensins are naturally occurring antimicrobial peptides secreted in the human body. Mammalian defensins are small, cysteine-rich, cationic peptides, generally consisting of 18–45 amino acids. The antimicrobial activity of defensins arises from their unique amino acid sequence, showing activity against both Gram-positive and Gram-negative bacteria, fungi and enveloped viruses. The use of antimicrobial peptides is rising due to their potential to control biofilm formation and kill microorganisms that are highly tolerant to antibiotics. In free-form, defensins are capable of destroying such microorganisms through numerous mechanisms mainly the *carpet*, the *toroidal* and the *Barrel-Stave* models. However, immobilization of antimicrobial peptides (AMPs) on surfaces with the help of coupling agents and spacers can improve the AMPs' lifespan and stability in the physiological environment leading to applications for medical devices and implants. Fundamental understanding of both free-form and surface-immobilized defensins is important to design more effective antimicrobial peptides and improve their performance in future developments.

Keywords: antimicrobial peptides, defensins, mammalian peptides, surface-immobilized antimicrobial peptides, surface-immobilized defensins

1. Introduction

The innate immune system is the first line of defence in human body and vertebrates. Defensins are naturally occurring antimicrobial peptides (AMPs) that are a part of the innate immune system, protecting the body against foreign microorganisms. Defensins are produced by the interaction of antigen-presenting microbial cells with pattern recognition receptors, such as toll-like receptors that are present on the membrane of numerous immune cells (i.e., macrophages, neutrophils and leukocytes [1]. Mammalian defensins are small, cysteine-rich, cationic peptides, generally consisting of 18–45 amino acids [2]. Next to being antimicrobial, defensins also serve as immune-stimulating agents.

When synthesized *in vivo*, defensins are initially produced as inactive precursor proteins (i.e., pro-defensins), which consist of the defensin and a pro-peptide. The pro-peptides are present to ensure delivery of defensins through the body without

IntechOpen

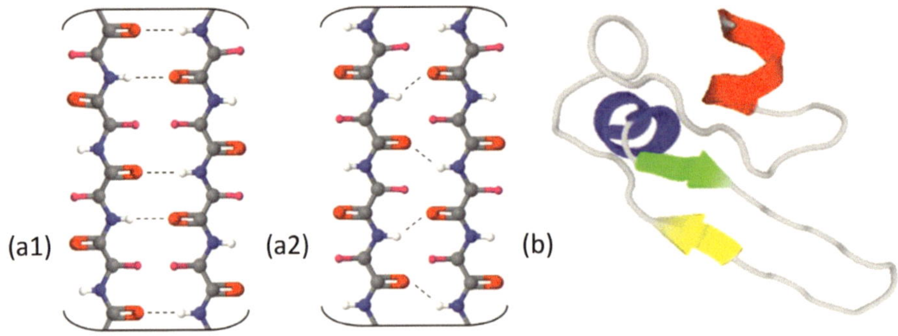

Figure 1.
Schematic of a (a1) parallel and (a2) antiparallel β-sheet structure, made with the use of JSME [12]. (b) Conformation of a β-defensin (hBD3), made with the use of PEP-FOLD3 [13]. Hydrogen bonding represented by dotted lines, hydrogen atoms are white, carbon atoms are grey, oxygen atoms are red, nitrogen atoms are blue and rest-groups are pink; not exact structure.

premature attachment of defensins to other microorganisms [3]. The pro-peptides inhibit premature attachment to other microorganisms by neutralizing the cationic charge of defensins. Also, the pro-peptides ensure subcellular localization (i.e., the location of where a protein resides in a cell) and folding of defensins into their characteristic conformation. Through proteolytic removal *in vivo* of the pro-peptides, the defensins are activated [4]. The reasoning behind the pro-peptides functioning as a folding assistant is based on the research performed on folding of defensins *in vitro* without the pro-peptide, which is found to be extremely difficult [5, 6].

α-Defensins are expressed by neutrophils and macrophages, that is, a type of white blood cell and cells that can engulf foreign particles, respectively. In general, these tend to have a broader antimicrobial activity, when compared to β-defensins, showing activity against both Gram-positive and Gram-negative bacteria, fungi and enveloped viruses [7]. Paneth cells also produce α-defensins, also known as crypticidins, which are involved in the reduction of bacteria present in the intestinal lumen. β-Defensins are primarily produced and released by epithelial cells and leukocytes, that is, a type of cell that lines the surfaces of your body and a type of blood cell that is made in the bone marrow, respectively. These are mainly active against Gram-negative bacteria and yeast; however, many also show antibacterial activity towards Gram-positive bacteria [8]. The pro-peptides of β-defensins are smaller than those of α-defensins.

Both α- and β-defensins form a triple-stranded antiparallel β-sheet structure that is stabilized by hydrogen and disulphide bonds; bond formations are schematically represented in **Figure 1**.[1] The position of the cysteines and intramolecular disulphide linkages determines the category of the defensin. The consensus of cysteine placement within the amino acid sequence for α-defensins follows C-X-C-X_4-C-X_9-X-X, and C-X_6-C-X_4-C-X_9-C-X_6-C-C for β-defensins [9]. When looking at the position of the disulphide linkages from cysteine in sequential order (denoted by C#), the disulphide bridges are formed between C1-C6, C2-C4 and C3-C5 for α-defensins and C1-C5, C2-C4 and C3-C6 for β-defensins [10], as shown in **Figures 2** and **3**, respectively. The disulphide bridges are important for holding the defensins in their three-dimensional structures. In addition, they contribute to the defensins chemotactic activity (i.e., movement or orientation of an organism or cell towards chemical stimulus) but when altered, only slightly affect their antimicrobial activity [11].

The adopted mechanisms of the interaction between defensins and the invading microorganism are not yet fully understood. However, disruption of

[1] All the figures shown in this chapter have been created by the authors.

DOI: http://dx.doi.org/10.5772/intechopen.85508

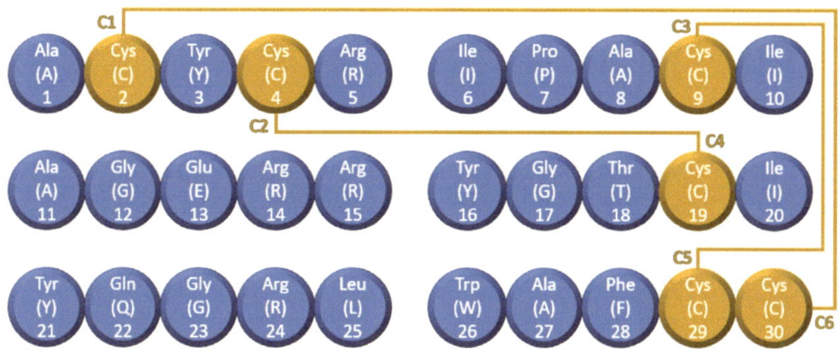

Figure 2.
Schematic of HNP1 (α-defensin) showing amino acid sequence and disulphide bridges.

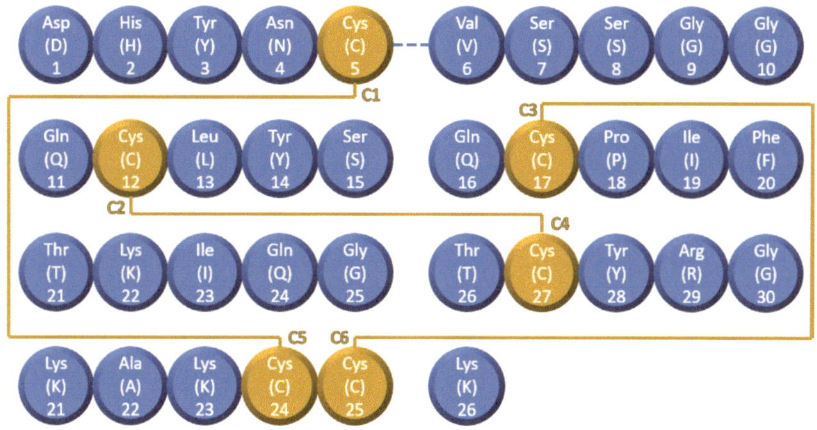

Figure 3.
Schematic of HBD1 (β-defensin) showing amino acid sequence and disulphide bridges.

the plasma membrane has been shown to be the leading cause of cell death in microbial species. The disruption caused by defensins depends on many factors, such as the polar topology, spatial separating of charges and hydrophobicity. These factors allow the attraction and subsequent interaction of defensins with the lipid bilayer of the bacterial membrane. Conversely, this interaction causes the defensins to insert themselves between the hydrophilic region of the plasma membrane and disrupt the bacterial membrane, utilizing numerous mechanisms. These include the introduction of channel-like pores and carpet-like membrane disruption, resulting in cell lysis. Simultaneously, the introduction of voltage-dependent channels in the bacterial membrane allows the influx of water and results in an increase of osmotic pressure that leads to the rupture of the membrane. On the other hand, some defensins move through bacterial cell walls, bind to target cells and disrupt normal metabolism, which may lead to apoptosis of the targeted cells [14, 15].

2. Mammalian defensins

The genomic organization and evolution of defensin genes of several vertebrate species have been studied [16]. The human genome encodes, at least, 35 different defensin peptides [17]. Most of the mammalian defensin genes are divided over three chromosomes, found in four different gene clusters (**Figure 4**). All the genes

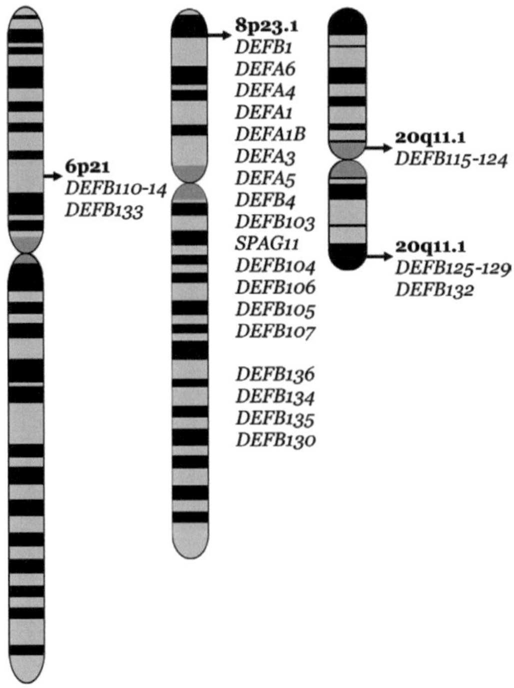

Figure 4.
Schematic of the genomic organization of defensins in three human chromosomes, showing chromosome 6, chromosome 8 and chromosome 20, respectively.

expressing α-defensins and several β-defensins are found in chromosome 8 (cluster p23.1); genes that express most of the remaining β-defensins are found in chromosome 6 (cluster p21) and chromosome 20 (cluster q11.1 and p13).

Four out of the six human α-defensins are found in neutrophils and other leukocytes, specifically, human neutrophil peptides (HNPs) 1–4 (i.e., DEFA1–4). The remaining two, human α-defensin 5 and 6 (HD5 and HD6, i.e., DEFA5 and DEFA6), are expressed by Paneth cells in the intestinal lumen [18]. Numerous β-defensins are found in the respiratory system, gastrointestinal tract and urogenital system. These are expressed by epithelial cells, namely, human BD1 (hBD1, i.e., DEFB1), hBD2 (DEFB4), hBD3 (DEFB103A) and hBD4 (DEFB104) [19].

Defensins are amphipathic (i.e., having both hydrophobic and hydrophilic groups) and, it has been demonstrated that they show the ability to form dimers and oligomers with toxin molecules [20, 21]. The initial electrostatic interaction is caused by the cationic charge of the peptide and the negatively charged outer membrane of the bacterial cell wall [22]. Bacterial membranes contain many negatively charged phospholipids, lipopolysaccharides or teichoic acid, while eukaryotic membranes contain neutral phospholipids and cholesterol [23]. This explains the destructive ability of defensins towards microbes but not host cells [24].

2.1 Antimicrobial mechanisms of defensins

The activity of defensins against microorganisms is determined by the interaction of the cationic molecules with the negatively charged acidic lipopolysaccharide or teichoic acid on Gram-positive and Gram-negative bacterial membranes. The antimicrobial activity of defensins in the body depends on different factors such as salt concentration and serum components [25]. In addition, the configuration of defensins plays an important role on the activity of these molecules. For example, helical antimicrobial peptides in solution show a typical helical amphipathic characteristic and they are either unstructured until they are in contact with the biological membrane or structured through disulphide bonds. Defensins are among the structured type of antimicrobial peptides and the hydrophilicity of the α- and β-defensins determines the extent of interaction between these molecules and the bacterial membrane [26].

As mentioned before, the antimicrobial mechanisms of defensins have not been fully understood. The formation of membrane pore or channel has been shown to be dependent on the membrane configuration. For example, the abundance of negatively charged phospholipids on the plasma membrane affects the concentration of peptides that are required to form a stretch or curvature on the outer layer of the bacterial membrane, and consequently lead to cell lysis [27]. Other events such as phospholipid reversal and penetration of peptides inside the cytoplasm on the inner side of the membrane leading to the loss of membrane composition and causing cellular inactivation have also been mentioned in the literature [28].

Most studies conducted on defensin mechanisms have been conducted on the α-helical structured peptides. These molecules interact with the membrane of the microorganisms. It is evident that defensins utilize membrane depolarization and permeation, against bacteria and yeast, as their most likely defence mechanisms. Defensins also aid the mobilization of T-cells and immature dendritic cells, which contribute in the activation of acquired immune responses that will trigger a long-lasting cellular response to a potential pathogen [29]. Most defensins provide their antimicrobial activity through interaction with cellular membranes. The pore-forming model describes the interaction between positively charged peptides and the negatively charged head of phospholipid groups of cellular membranes operates once a critical concentration is reached. This will cause the self-aggregation and perpendicular insertion of peptides inside the membrane leading to production of lined transmembrane pores, resulting in the disruption of ionic and proton gradients. The second model causes the formation of channels where the strain of peptide at the critical concentration, induces the inward curving of the membrane, creating dome-shaped channels lined with phospholipid-head groups and peptides [30].

The mechanisms in which the AMPs destroy the bacterial cells have been studied for decades. AMPs that present their secondary structure in the form of α-helix present their destructive effects on a bacterial cell based on three different documented mechanisms. First is the *Barrel-Stave model*, which is the most studied mechanism, where the peptide disrupts the membrane by exposing its hydrophobic site to the lipids in the membrane bilayer. This will force the membrane to undergo conformational changes by forming a pore on the surface of the membrane [31, 32]. The second mechanism is the *toroidal model* where the peptide and the lipid of the membrane bilayer integrate upon interaction and form torus pores leading to the death of bacterial cells. Cell death caused by these mechanisms leads to loss of compositional specificity, leakage of critical metabolites or depolarization of the membrane due to an increased rate of phospholipid reversal [33, 34]. The third mechanism is the *carpet model* where the concentration of AMP is related to the amount of interaction with the bacterial membrane. In this model, the peptides surround the cell membrane by attaching to the phosphide group of the membrane bilayer where they disrupt the curved anionic membrane and dissolve it, killing the bacteria [35, 36].

2.2 Antimicrobial activity evaluation methods

The evaluation of antimicrobial activity is widely used in the field of drug discovery as well as epidemiology and therapeutic prediction [37]. Since the low density of peptides will ultimately result in preservation of membrane integrity and adjustment of the membrane, maintaining a higher concentration of peptides is crucial. This will result in imposed curvature strain on bacterial membrane [38]. Therefore, measurement of the minimum inhibitory concentration (MIC) is an important initial step for the confirmation of antimicrobial activity. MIC refers to the lowest concentration of compound needed for observable inhibition of bacterial cell growth. In addition, other measurements include minimal bactericidal concentration (MBC) which refers to the minimum concentration of the reagent that can cause bacterial death, the haemolytic activity (HC) that refers to the ability to break down red blood cells (i.e.,) to find the inhibitory effect of AMPs on normal mammalian cells and IC_{50}, which refers to the half-maximal inhibitory concentration [39]. Since defensins are found in more diluted concentrations in extracellular environment than in a local environment, the measurement of the interactions between peptides and bacteria *in situ* (e.g., using a mice model) is mandatory [40].

Colorimetric assays are used to determine the concentration of peptides (MIC) where the absorbance is usually measured at 750 nm using a UV-visible spectrophotometer. The assays are performed in 96-well microtiter plates and a series of antimicrobial peptide dilutions are added to the bacterial cells immersed in growth medium. The MIC is measured based on the growth after incubation for a defined period of time (16–20 h) [41]. This methodology measures the colour of the dilutions, which is directly proportional to the number of cells in each well. The absorbance is measured with the help of a microplate reader counting the number of cells killed by the antimicrobial peptide. This methodology provides information about the susceptibility of the microorganism to the peptide [42], but it only applies to aerobic bacteria [43].

Electron microscopy is used to visualize the interaction of peptides with the phospholipid bilayer of bacterial membrane. Simultaneously, fluorescent dyes are used to observe the ability of peptides to permeate and penetrate the membrane of bacterial cells. Fluorescence spectroscopy is used to study the insertion of fluorophores into a membrane as well as provide information about the rate of peptide penetration into the membrane with the help of surface plasmon resonance. Fluorescence quenching can also be used to gain an insight into the depth of peptide penetration inside the cells [2].

In addition, both circular dichroism (CD) spectroscopy [44] and nuclear magnetic resonance (NMR) spectroscopy are used to measure the orientation and secondary structure of an antimicrobial peptide, when bound to a lipid bilayer. However, only NMR is used to measure the penetration of antimicrobial peptides into lipid bilayers in a relevant liquid-crystalline state [45]. The use of CD can distinguish between the randomly coiled, α-helical and β-sheet structures since they show wavelength-dependent differences in the absorption of the right and left circularly polarized light. Using oriented CD can also provide information about the orientation of peptide upon insertion into the membrane. This is dependent on the concentration of these peptides, the nature of the lipid and the extent of hydration [2].

In order to study the peptide configuration, both solution and solid-state NMR can be employed. For solution NMR, a mixture of peptide-detergent micelle is needed for the stabilization of the peptide in water and to overcome insufficient resolution and low (signal/noise) ratio [46]. The most recognized model membrane system used in solution NMR is the dodecylphosphocholine (DPC). This method is used for studying the interaction of peptides with the lipid bilayer since it has the ability to rotate freely in solution and mimic anisotropic environments of lipid membranes [47]. In addition, solid-state NMR is a premium technique to use for the analysis of immobile peptides that are difficult to analyse with crystallography or solution NMR. They can be used without a need for major peptide modification to determine the structure of membrane proteins in the phospholipid bilayer. This technique offers the examination of the structure and motion for the peptide-lipid interactions in physiologically relevant conditions and produces sharp resonance lines due to macroscopic alignment [48].

In addition, to provide the dynamic interaction between the defensins and the lipid bilayer, using a monolayer mimicking the structure of bacterial membranes can also be used. To observe the interaction between the two, the sum frequency generation spectroscopy (SFG) is employed. This methodology can provide information about gas-liquid, solid-liquid and liquid-liquid interactions and is able to detect the biomolecule orientation and adsorption in sub-micron quantities. This technique uses a pulsed narrow band visible laser beam spatially and temporally overlapped with a broadband infrared laser and measures the incident beam produced from the surface [49]. The vibrational spectrum of C=O groups from the amide backbone group can provide the secondary structure of the peptides. Consequently, vibrations from acyl chains of the lipid bilayer can provide important information about the interaction of peptide with the membrane as well as information about the molecular structure of the peptide, without the use of vesicles and labels to complicate the process of analysis [50].

3. Antimicrobial activity of defensins in free-form

There are two types of AMPs: first AMPs that show activity towards both bacterial and mammalian cells; second, AMPs that show only activity towards the bacterial cells. Most linear cationic AMPs are unordered in aqueous solution. The balance between the positively charged and hydrophobic amino acids in cationic AMPs permits the amphipathic adaptation of these molecules in solution. This allows the interaction of AMPs with the negatively charged bacterial membrane and the subsequent penetration of these molecules inside the lipid membrane [51]. The interaction of these peptides with the bacterial membrane is increased due to their high inside-negative transmembrane potential. On the other hand, normal eukaryotic cells have a net neutral charge across their membrane bilayer, and they have reached a zwitterionic (overall neutral) point. This insight can partly explain the attraction of AMPs to prokaryotic cells and the relatively weak attraction of these molecules towards eukaryotic cells [38]. The amphipathic characteristics of defensins aid the adaptation of a folded confirmation for these molecules in both hydrophobic and hydrophilic environments [44]. These molecules are often difficult to stabilize and show poor bioavailability due to their many different cleavage points that provide susceptibility to enzyme degradation and their linear form which leaves their two ends exposed [52].

α-Defensins (DEFA1–4) are produced by endoplasmic reticulum of the bone marrow shown in **Figure 5**, and the highest concentration of defensins is found in granules (i.e., leukocyte storage organelles). Pre-pro-defensins (light purple circles in **Figure 5**), however, consist of 94 amino acids. To produce pro-defensins (dark purple circles in **Figure 5**), 19 amino acids are removed from the N-terminus of pre-pro-defensins. Further proteolysis from the N-terminal side of the sequence of amino acids results in the production of mature defensins (blue circles in **Figure 5**). The prepared defensins are encapsulated into vesicles and fused with phagocytic vacuoles. The introduction of a pathogen into cells will then result in the recognition and engulfing of such molecules by phagocytic leukocytes with the aid of defensins and the subsequent death of the pathogenic organism. Although some α-defensins are produced in the bone marrow, some (DEFA5-6) are also produced

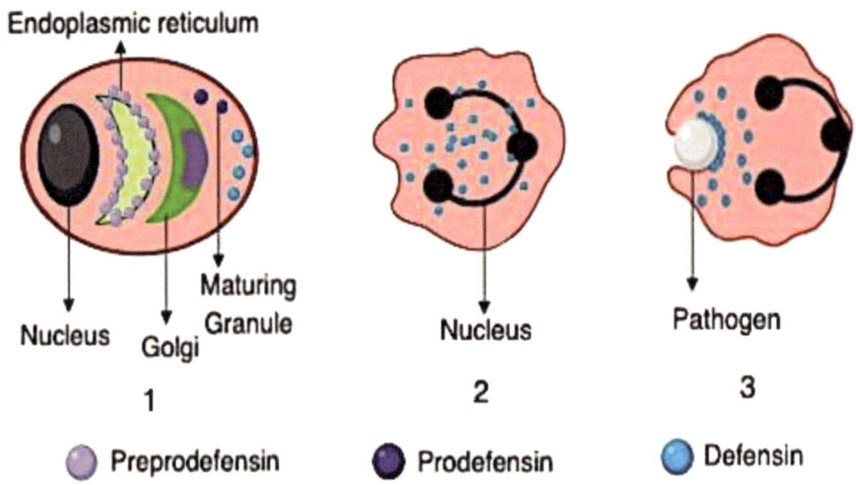

Figure 5.
Schematic diagram of production of α-defensins and the phagocytosis of pathogens with the aid of defensins: (1) bone-marrow promyelocyte; (2) neutrophil; (3) phagocytosis.

in the Paneth cells of the intestines [53]. β-Defensins (DEFB1–4) on the other hand are mainly produced in the epithelial cells. These regions have the highest concentration of defensins due to the higher susceptibility to a pathogenic attack, which renders them weaker and subsequently in need of the immunity provided by defensins [17].

β-Defensins mature into a secreted peptide after pre-pro-peptide state. This mature peptide has six cystine residues connected with intramolecular disulphide bonds. The connectivity of these disulphide bonds as well as the number of residues are factors differentiating α- and β-defensins [54]. The importance of a balance between the hydrophobicity of the defensins and their net positive charge is highlighted in the activity of these peptides towards the bacterial membrane [55]. The crystal formation of defensins usually results in the production of a dimeric structure. Crystallographic studies of α- and β-defensins show that free-form α-defensins have three intramolecular disulphide bonds (Paneth cell defensins and innate immunity of the small bowel) and they form a dimeric structure with six β-sheets [56]. The monomeric structure of β-defensins consists of three β-sheet folds as well as a helical N-terminus [16]. Nevertheless, defensins undergo conformational changes when introduced to a bacterial membrane. Targeted approaches on a specific protein may render the AMPs useless against the bacteria due to bacterial resistance and genetic changes to its conformation. Therefore, it is important that defensins work as non-specific agents on the membrane of the bacterial cells. Bacteria resistant to antimicrobial peptide activity usually display enzymatic covalent modification on their membrane, which reduces their negative charge.

α- and β-defensins are generally known for their cluster of positively charged amino acid residues. Although the sequence of amino acids is highly variable in defensins, their cysteine residues and their framework are highly conserved [17]. DEFA1-3 and human β-defensins are stored as mature peptides, whereas DEFA5 is stored in its pro-peptide form [57, 58]. The dimerization of β-defensins in solution is shown to be the functional structure for recognition of microorganisms [16]. However, DEFA1 takes a non-dimeric structure in solution, forming a voltage-dependent channel in the planar lipid bilayer [59]. This specific example also shows that the interaction of DEFA1 with the membrane phospholipid depends on the presence of anionic phospholipids on the membrane. On the other hand, DEFA2, an α-defensin similar to DEFA1 (lack of alanine at position 1), requires the assembly of approximately two dimers for the formation of pores in uni-lamellar vesicles [60]. DEFA1 and DEFA2 are the potent forms of α-defensins and DEFA3, differing by only an additional amino acid at the N terminus, is known to be less active against *C. albicans* [61]. The dome-shaped, three-dimensional structure of amphiphilic defensins has been hypothesized to have the N- and C-termini at the two ends and the hydrophobic section having an amphiphilic structure at the lowest portion. The functional diversity of defensins and their potency mainly depend on the N- and C-terminal residues [26]. A study showed that arginine-rich cationic defensins provide a higher spectrum of antimicrobial activity due to their higher cathodal electrophoretic mobility [62].

The concentration of defensins in epithelial cells averages 10–100 μg ml^{-1} although the uneven distribution of these molecules results in a higher local concentration [17]. Defensins attack Gram-positive and Gram-negative bacteria with the same mechanisms; however, the attack is on the cell wall and the outer membrane of the bacteria, respectively. Accumulation of cationic peptides close to the negatively charged surface in Gram-negative bacteria may lead to binding and crossing of defensins into the cell via a charge-exchange mechanism by competing with Ca^{2+} and Mg^{2+} bound to lipopolysaccharides [63]. Although the porous surface of Gram-positive bacteria allows ease of movement for the defensins [64], safe passage of defensins through the cell wall allows via the same mechanisms to attack the membrane bilayer. Generally, defensins are known to interact with lipopolysaccharides, polysaccharides and phospholipids of Gram-negative, Gram-positive and bacterial membrane bilayer, respectively [65].

Defensins are active against bacteria at a concentration of 1–10 μg in optimal conditions such as low ionic strength conditions, low concentrations of proteins or other substances interfering with this activity. However, cellular conditions are harsh and the salt as well as protein concentrations inhibit defensins' antimicrobial activity depending on the sequence and bacterial target of the defensin [66]. In addition, higher concentrations of defensins have been shown to have toxic effect towards mammalian cells, specifically lung tissue [67]. Permeabilization of defensins renders the production and synthesis of DNA and their subsequent RNA and protein. The extent of interaction of defensins with the bacteria depends on the amino acid backbone and the flexibility of this chain allowing for the presence of potential spatial interactions with the head of phospholipid groups in the membrane [68].

Defensins provide their antimicrobial activity by the creation of pores or membrane disruption, which both lead to the release of cellular contents [69]. Other regulatory factors such as wound closures [70], fibroblast proliferation and chemotaxis of T-cells and dendritic cells can also be mentioned as the activity of defensins. Defensins also have a modulatory effect on the production of cytokines [71]. Other activities of defensins include regulation of immune and inflammatory regions by providing chemotactic activity for monocytes, T-cells and dendritic cells. In addition, nanomolar concentrations of defensins can also activate nifedipine-sensitive calcium channels of mammalian cells, reducing the electrical conduction [72].

Overall, defensins are important molecules for providing a cascade of antimicrobial activity in the human body in defence against pathogens and they are mainly concentrated in regions that are more prone to pathogenic attack.

4. Defensin-like peptides immobilized on material surfaces

Bacterial adhesion and colonization resulting in biofilm formation on the surface of biomaterials are responsible for most medical device-associated infections and malfunctions [73, 74]. The magnitude of this problem continues to pose a significant problem to health care providers, often resulting in major negative medical and economic consequences [75, 76]. Many of the currently used medical devices are amenable to modification, either by functionalization or coating of the surface of the device, making it possible to combat inflammation and reduce the risk of infection [77].

In general, device-associated infections have been treated by developing material surfaces containing antibiotics and biocides [78]. The use of this approach comes with the risk of cytotoxicity, raising concern as a potential threat to human and environmental health. An alternative to this is the use of antifouling coatings, making it possible to prevent attachment of bacteria, proteins and other microorganisms. This more passive approach makes it possible to prevent antibiotic resistance and leaching of cytotoxic biocides but is not capable of killing already adherent bacteria and makes it hard to avoid infection completely [79]. Therefore, a combination of antifouling and antimicrobial properties is favourable to combat biofilm formation and further reduce the risk of infection [80, 81].

Defensin-like peptides (i.e., antimicrobial peptides, AMPs) exhibit a combination of antimicrobial and antifouling properties, which is why AMPs have received significant attention as an alternative to conventional biocides and antibiotics, showing the ability to overcome and combat medical device-associated infections. However, so far, there is little success in the development of AMPs for therapeutic applications, with only a few AMPs that have been approved for medical use at their initial introduction. After the unsuccessful introduction of the 'first-generation' AMPs, research has been performed on reducing cytotoxicity. It was found that AMPs were increasingly more toxic when having hydrophobic characteristics, sufficient enough to interact with neutrally charged eukaryotic cell membranes [82]. By replacing or interrupting these hydrophobic regions of AMPs, cytotoxicity was reduced and only showed a slight decrease in antimicrobial activity, providing the amphipathic characteristics were maintained [83]. Also, developing narrow-spectra AMPs would decrease the required concentration needed to combat pathogens and additionally prevent the cytotoxic activity towards eukaryotic cells [84].

In order to further reduce the cytotoxicity, immobilization of AMPs onto material surfaces is a potential approach to reduce the concentration needed when in free-form and will also increase their half-life time. The half-life time of AMPs is found to be based on the rate of protease digestion or related to peptide aggregation [85]. In order to compensate for their relatively short half-life time, increased concentrations of AMPs are used; but, this results in increased cytotoxicity and has limited the use of AMPs. Other efforts to increase the half-life time of peptides have been focused on using substitutes for L-amino acids, because unprotected peptides are more rapidly metabolized (i.e., broken down for nutrition) [86]. By substitution of L-amino acids by D-amino acids, the introduction of unnatural β- and γ-peptide bonds and modifications of the N- or C-terminus, it is possible to increase the stability of AMPs [87–89].

Overall, it is of great importance to improve the stability of AMPs against degradative mechanisms *in vivo* and increase the bond stability between the AMPs and materials in order to develop almost non-cytotoxic and long-lasting antimicrobial surfaces [90, 91]. Therefore, by limiting the necessity of implant removal due to AMP inactivity caused by degradative mechanisms, patient compliance can be increased.

4.1 AMP immobilization methods

It is possible to immobilize AMPs through two main pathways, either physically or chemically. A popular physical method is layer-by-layer assembly, in which AMPs are 'sandwiched' between two polyionic polymers, making it possible to integrate a controllable loading of AMPs [92, 93]. However, the interspersed AMPs within the polyionic polymer layers are not able to interact with the surrounding environment, and they will need to diffuse outwards to utilize their antimicrobial activity. Therefore, with the use of physical immobilization, it is difficult to fully utilize the potency of AMPs. Covalent-based immobilization of AMPs has significant advantages, in comparison to physical immobilization, such as the formation of more stable bonds and thus improvement of their relatively short half-life time, while also minimizing the possibility of leaching of AMPs [90, 94].

Furthermore, the material surface is also of great importance to the stability of AMPs. A polymer surface (i.e., dibromomaleimide polymer substrate) prepared by chemical vapour deposition (CVD) and functionalized with AMPs showed a better antimicrobial stability when compared to a self-assembled monolayer (SAM). When exposed to air, this polymer showed slower detachment of bound AMPs when compared to SAM [95]. Additionally, orientation of AMPs was also retained. These results show the importance of using a non-degradable material surface to improve bond stability of AMPs. Next to that, the surface morphology needs to be well defined as an undefined surface could lead to inhibition of attachment of biological molecules. Also, depending on the density of the functional groups that are present or able to be induced, the number of peptides on the material surface can vary significantly. However, even though AMP concentration is of importance, it does not appear to be the most critical criterion for the improvement of antimicrobial activity [96, 97].

The antimicrobial activity of covalent-based surface-immobilized AMPs is seen to be mainly dependent on the used coupling strategy, spacer specifications and peptide orientation and concentration [98]. However, there are certain limits to improving the antimicrobial activity by increasing the AMP concentration, due to factors such as coupling conditions and steric hindrance (i.e., repulsive forces originating from overlapping electron clouds of neighbouring molecules). Next to that, some microorganisms are found to be insensitive to any further increase after reaching a certain limit in AMP surface density or exposure time [99].

It is demonstrated that direct-immobilized AMPs also show antimicrobial activity without the use of a spacer. Nevertheless, most potent developments make use of spacers, the length (i.e., the distance between the material surface and peptide) of which is shown to significantly influence the activity of surface-immobilized AMPs, when compared to the AMP surface density [90]. Conversely, the possibility of chain cleavage of the spacer due to polymer degradation reactions could lead to the release of immobilized AMPs. This could be minimized with the use of stabilized polymer spacers [59]. The increased activity seen with the use of spacers is a result of improved mobility of AMPs, increasing probability of membrane permeabilization and subsequent cell death. However, this would only be a correct hypothesis assuming that AMPs demonstrate a similar mode of action to that of their free-form counterparts. According to the literature, there are also AMPs that will depolarize the cytoplasmic membrane and disrupt the electron transport, which subsequently would lead to partial membrane permeabilization and thus cell death [100, 101].

The difficulty of these developments is the lack of comparable information, since most of the observed reactivity and antimicrobial activity are found to be specific to the reactant environment, type of materials and AMPs. Fully understanding these structure-function relationships is important to clarify and improve the performance of surface-immobilized AMPs in future developments.

4.1.1 Orientation of direct immobilized AMPs

When a peptide is directly bonded to the material surface, as displayed in **Figure 6**, the immobilized terminus determines its orientation. When α-helical peptides are immobilized at their N-terminus, they will point perpendicular to the material surface (i.e., orientate upstanding), but when immobilized at their C-terminus, they will take on a laying-down orientation. In general, when an AMP has its N-terminus bonded to the surface, it has been shown to have a lower MIC when compared to C-terminus and N-side-chain-immobilized AMPs [102]. The relatively high MIC of C-terminus-immobilized AMPs is likely related to the inhibition of membrane interaction [51]. However, according to the literature reporting the antimicrobial activity dependence of the orientation difference between N-terminus and C-terminus-immobilized peptides, there is also a cell-dependent potency, which indicates that the mode of action of AMPs is not only dependent on their own characteristics but also on the characteristics of the targeted microorganism [103]. Another important parameter is the position of the cationic amino acids. When they were closer to the bonding site, an increased antimicrobial activity was observed, while, when they were positioned in the middle or closer to the N-terminus, a decrease of the antimicrobial activity was observed [100].

However, as shown in **Figure 7**, the orientation of AMPs changes after their initial interaction with bacteria (i.e., the immobilized α-helices bind to the anionic lipid bilayer). Since immobilized AMPs cannot follow the *barrel-stave* or *toroidal model* due to limited mobility, it is suggested that the charge-charge interaction plays a dominant role in the elimination of bacteria [104, 105].

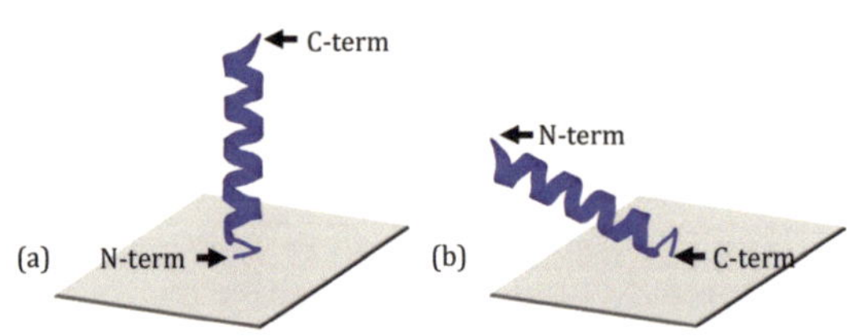

Figure 6.
Schematic of (a) N-terminus and (b) C-terminus surface-immobilized α-helical peptide; not drawn to scale or exact structure.

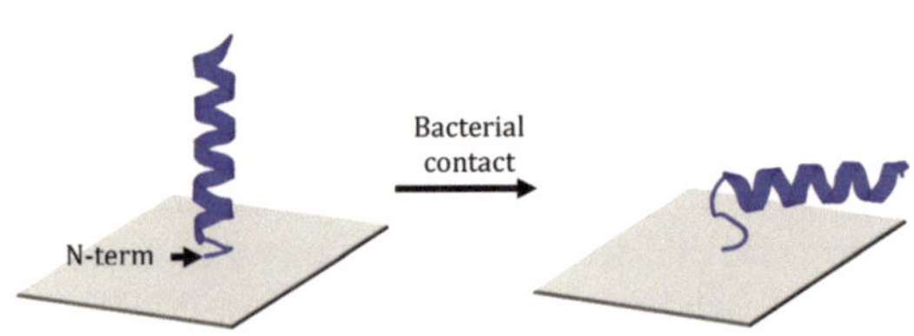

Figure 7.
Schematic of orientation change of N-terminus-immobilized α-helical peptide before (left) and after (right) bacterial contact; not drawn to scale or exact structure.

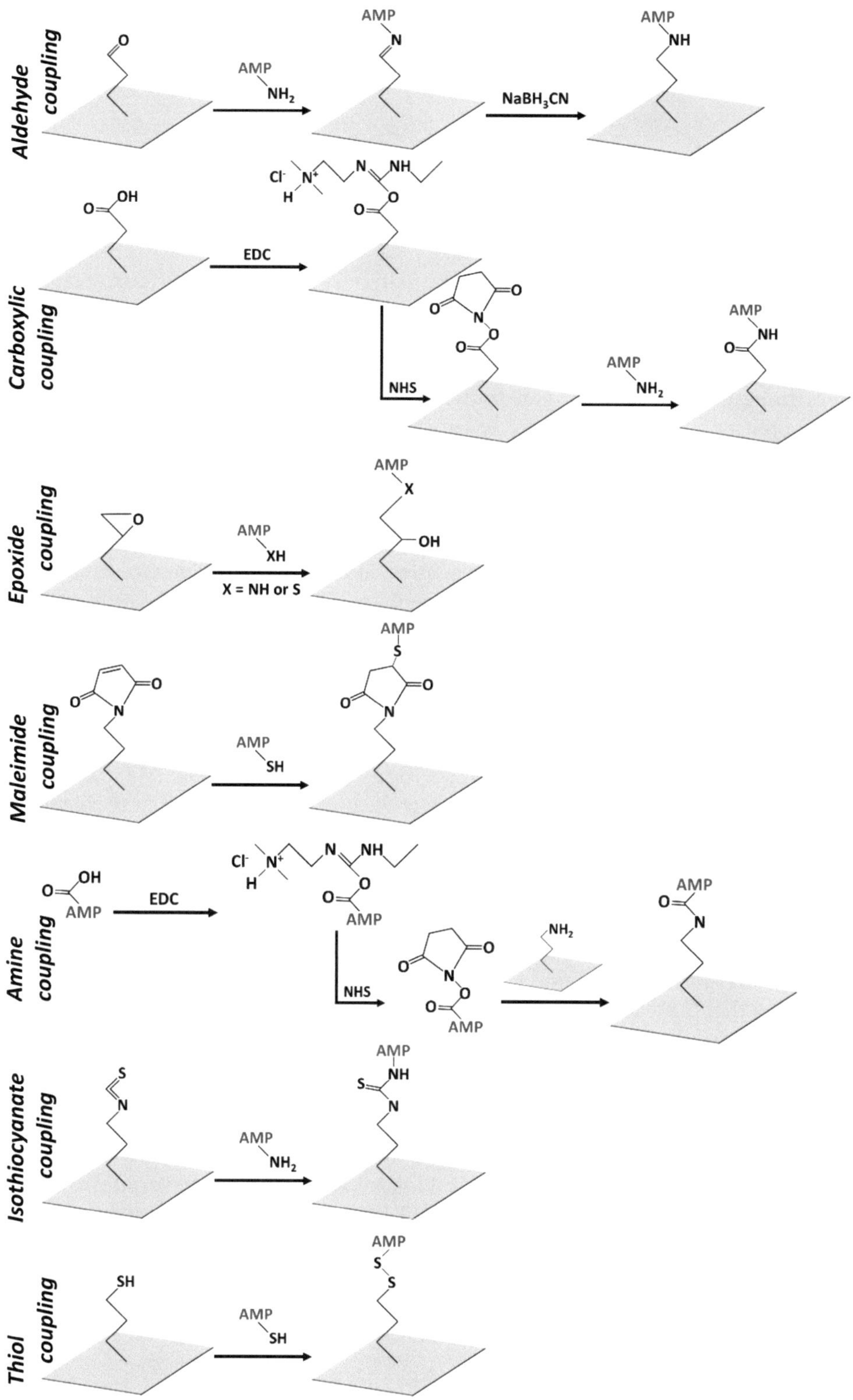

Figure 8.
Immobilization of AMPs onto numerous functional groups, using different coupling methods; not drawn to scale or exact structure.

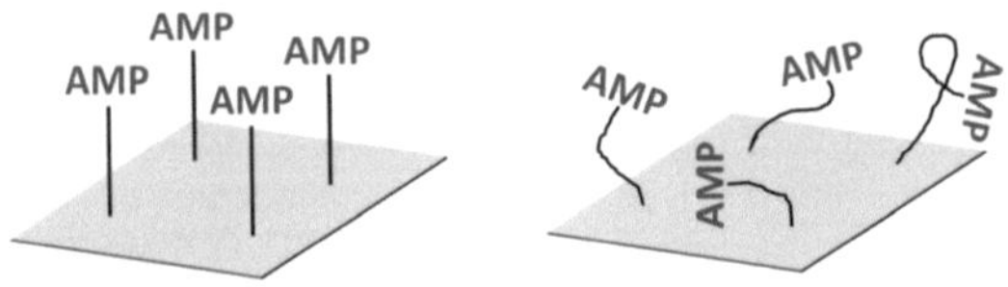

Figure 9.
Schematic of spacer-incorporated immobilized AMPs, showing stiff spacers (left) and flexible spacers (right); not drawn to scale or exact structure.

Additionally, when functionalizing the material's surface with different reactive groups as seen in **Figure 8**, the orientation of immobilized AMPs can be controlled using chemo-selective (i.e., directed immobilization) coupling reactions. In general, 1-ethyl-3-(3-dimethylaminopropyl)carbodiimide (EDC) and N-hydroxysuccinimide (NHS) are found to be widely used as activating agents for the coupling of peptides to numerous functional groups by forming amides, which inhibit the formation of carboxylic salts with an amine [98, 106].

4.1.2 Spacer-incorporated immobilized AMPs

Although orientation can be partly controlled with the use of chemo-selective coupling agents, they often do not add enough mobility to increase the probability of membrane permeabilization or polarization. In order to improve the mobility and subsequent activity of AMPs, the use of spacers is found to be useful. As shown in **Figure 9**, there are two types of spacers, stiff and flexible spacers. Stiff spacers (e.g., polyvinyl chloride or polymethyl methacrylate) allow the increase of reach and thus might allow membrane permeabilization, but will restrict sideways mobility by keeping the AMPs pointing in a specific orientation [98]. Flexible spacers such as polyethylene glycol (PEG) also allow the increase of reach and are able to allow sideways mobility; however, orientation cannot be determined due to their flexible chains [107].

However, even if the peptide was linked to a PEG spacer, the random orientation of an immobilized AMP through its C-terminus is found to result in the loss of antimicrobial activity [97]. However, the oriented immobilization of the same AMP through its N-terminus is found to restore the antimicrobial activity. It has also been suggested that the water-swelling property of PEG aids in maintaining the activity of immobilized peptides [108]. Additionally, in the absence of the PEG spacer, the AMP did not show antimicrobial activity. Nevertheless, the utilization of PEG as a spacer is found to present numerous advantages, as it can create non-adhesive surfaces due to its non-fouling characteristics (i.e., inhibition of microorganisms binding to the material surface) [108]. Lastly, the solubility and stability of peptides, against protease digestion or peptide aggregation, can be improved with the use of a spacer [107, 109, 110]. These previously mentioned factors show yet again the complexity of the factors on influencing the antimicrobial activity of immobilized AMPs.

5. Conclusions

AMPs are demonstrated to show antimicrobial activity at relatively low concentrations, without damaging mammalian cells, being able to utilize several mechanisms against numerous microorganisms similar to defensins in the human body. Immobilization of AMPs improves their lifespan, preserves the mode of action and does not seem to influence the mechanism on the biological level; however, they do show reduced antimicrobial activity upon immobilization to material surfaces. Nonetheless, this reduced activity can be partly restored with the use of chemo-selective coupling agents and the incorporation of spacers. Whether degradation of the material and coupling agents is desirable or not, understanding the kinetics is of great importance as the decrease of structural integrity and/or release of particles (i.e., molecules, debris, etc.) might result in an adverse biological reaction. Nonetheless, there are significant indications that AMPs are suitable candidates to replace conventional biocides and antibiotics. In addition, they can be utilized to develop the next generation of antimicrobial surfaces as coatings for medical devices and implants.

Conflict of interest

The authors certify that they have NO affiliations with or involvement in any organization or entity with any financial interest (such as honoraria; educational grants; participation in speakers' bureaus; membership, or other equity interest; and expert testimony or patent-licensing arrangements) in the subject matter or materials discussed in this chapter.

Author details

Marc Bruggeman, Hanieh Ijakipour and Artemis Stamboulis*
Biomaterials Group, School of Metallurgy and Materials,
University of Birmingham, Birmingham, United Kingdom

*Address all correspondence to: a.stamboulis@bham.ac.uk

References

[1] Hazlett L, Wu M. Defensins in innate immunity. Cell and Tissue Research. 2011;**343**(1):175-188

[2] Epand RM, Vogel HJ. Diversity of antimicrobial peptides and their mechanisms of action. Biochimica et Biophysica Acta (BBA) - Biomembranes. 1999;**1462**(1-2):11-28

[3] Michaelson D, Rayner J, Couto M, Ganz T. Cationic defensins arise from charge-neutralized propeptides: A mechanism for avoiding leukocyte autocytotoxicity? Journal of Leukocyte Biology. 1992;**51**(6):634-639

[4] Wu Z, Li X, Ericksen B, de Leeuw E, Zou G, Zeng P, et al. Impact of pro segments on the folding and function of human neutrophil α-defensins. Journal of Molecular Biology. 2007;**368**(2):537-549

[5] Raj PA, Antonyraj KJ, Karunakaran T. Large-scale synthesis and functional elements for the antimicrobial activity of defensins. The Biochemical Journal. 2000;**347**(3):633-641

[6] Wu Z, Powell R, Lu W. Productive folding of human neutrophil α-defensins in vitro without the pro-peptide. Journal of the American Chemical Society. 2003;**125**(9):2402-2403

[7] Lehrer RI, Ganz T. Endogenous vertebrate antibiotics: Defensins, protegrins, and other cysteine-rich antimicrobial peptides. Annals of the New York Academy of Sciences. 1996;**797**(1):228-239

[8] Harder J, Bartels J, Christophers E, Schröder J-M. A peptide antibiotic from human skin. Nature. 1997;**387**(6636):861

[9] Dorin JR, McHugh BJ, Cox SL, Davidson DJ. Mammalian antimicrobial peptides; defensins and cathelicidins. In: Molecular Medical Microbiology. 2nd ed. Elsevier; 2015. pp. 539-565

[10] Zhao B-C, Lin H-C, Yang D, Ye X, Li Z-G. Disulfide bridges in defensins. Current Topics in Medicinal Chemistry. 2016;**16**(2):206-219

[11] Wu Z, Hoover DM, Yang D, Boulegue C, Santamaria F, Oppenheim JJ, et al. Engineering disulfide bridges to dissect antimicrobial and chemotactic activities of human β-defensin 3. Proceedings of the National Academy of Sciences. 2003;**100**(15):8880-8885

[12] Bienfait B, Ertl P. JSME: A free molecule editor in JavaScript. Journal of Cheminformatics. 2013;**5**(1):24

[13] Lamiable A, Thévenet P, Rey J, Vavrusa M, Derreumaux P, Tufféry P. PEP-FOLD3: Faster de novo structure prediction for linear peptides in solution and in complex. Nucleic Acids Research. 2016;**44**(W1):W449-W454

[14] White SH, Wimley WC, Selsted ME. Structure, function, and membrane integration of defensins. Current Opinion in Structural Biology. 1995;**5**(4):521-527

[15] Zhang L-J, Gallo RL. Antimicrobial peptides. Current Biology. 2016;**26**(1):R14-R19

[16] Hoover DM, Chertov O, Lubkowski J. The structure of human β-defensin-1 new insights into structural properties of β-defensins. The Journal of Biological Chemistry. 2001;**276**(42):39021-39026

[17] Ganz T. Defensins: Antimicrobial peptides of innate immunity. Nature Reviews. Immunology. 2003;**3**(9):710

[18] Mallow EB, Harris A, Salzman N, Russell JP, DeBerardinis RJ, Ruchelli E, et al. Human enteric defensins gene structure and developmental

expression. The Journal of Biological Chemistry. 1996;**271**(8):4038-4045

[19] Pazgier M, Hoover DM, Yang D, Lu W, Lubkowski J. Human β-defensins. Cellular and Molecular Life Sciences. 2006;**63**(11):1294-1313

[20] Andreu D, Rivas L. Animal antimicrobial peptides: An overview. Peptide Science. 1998;**47**(6):415-433

[21] Zhao L, Ericksen B, Wu X, Zhan C, Yuan W, Li X, et al. Invariant Gly residue is important for α-defensin folding, dimerization, and function a case study of the human neutrophil α-defensin HNP1. The Journal of Biological Chemistry. 2012;**287**(23):18900-18912

[22] Hancock REW, Chapple DS. Peptide antibiotics. Antimicrobial Agents and Chemotherapy. 1999;**43**(6):1317-1323

[23] Matsuzaki K. Why and how are peptide-lipid interactions utilized for self-defense? Magainins and tachyplesins as archetypes. Biochimica et Biophysica Acta (BBA): Biomembranes. 1999;**1462**(1-2):1-10

[24] Zasloff M. Antimicrobial peptides of multicellular organisms. Nature. 2002;**415**(6870):389

[25] Peschel A. How do bacteria resist human antimicrobial peptides? Trends in Microbiology. 2002;**10**(4):179-186

[26] Raj PA, Dentino AR. Current status of defensins and their role in innate and adaptive immunity. FEMS Microbiology Letters. 2002;**206**(1):9-18

[27] Sahl H-G, Pag U, Bonness S, Wagner S, Antcheva N, Tossi A. Mammalian defensins: Structures and mechanism of antibiotic activity. Journal of Leukocyte Biology. 2005;**77**(4):466-475

[28] Matsuzaki K, Murase O, Fujii N, Miyajima K. An antimicrobial peptide, magainin 2, induced rapid flip-flop of phospholipids coupled with pore formation and peptide translocation. Biochemistry. 1996;**35**(35):11361-11368

[29] Ghosh SK, Gerken TA, Schneider KM, Feng Z, McCormick TS, Weinberg A. Quantification of human β-defensin-2 and-3 in body fluids: Application for studies of innate immunity. Clinical Chemistry. 2007;**53**(4):757-765

[30] Kraszewska J, Beckett MC, James TC, Bond U. Comparative analysis of the antimicrobial activities of plant defensin-like and ultrashort peptides against food-spoiling bacteria. Applied and Environmental Microbiology. 2016. AEM--00558

[31] Sani M-A, Separovic F. How membrane-active peptides get into lipid membranes. Accounts of Chemical Research. 2016;**49**(6):1130-1138

[32] Yang L, Harroun TA, Weiss TM, Ding L, Huang HW. Barrel-stave model or toroidal model? A case study on melittin pores. Biophysical Journal. 2001;**81**(3):1475-1485

[33] Balhara V, Schmidt R, Gorr S-U, DeWolf C. Membrane selectivity and biophysical studies of the antimicrobial peptide GL13K. Biochimica et Biophysica Acta (BBA): Biomembranes. 2013;**1828**(9):2193-2203

[34] Allende D, Simon SA, McIntosh TJ. Melittin-induced bilayer leakage depends on lipid material properties: Evidence for toroidal pores. Biophysical Journal. 2005;**88**(3):1828-1837

[35] Ghosh C, Haldar J. Membrane-active small molecules: Designs inspired by antimicrobial peptides. ChemMedChem. 2015;**10**(10):1606-1624

[36] Pouny Y, Rapaport D, Mor A, Nicolas P, Shai Y. Interaction of antimicrobial dermaseptin and its

fluorescently labeled analogs with phospholipid membranes. Biochemistry. 1992;**31**(49):12416-12423

[37] Balouiri M, Sadiki M, Ibnsouda SK. Methods for in vitro evaluating antimicrobial activity: A review. Journal of Pharmaceutical Analysis. 2016;**6**(2):71-79

[38] Bechinger B, Gorr S-U. Antimicrobial peptides: Mechanisms of action and resistance. Journal of Dental Research. 2017;**96**(3):254-260

[39] Munoz-Bonilla A, Fernández-Garcia M. Polymeric materials with antimicrobial activity. Progress in Polymer Science. 2012;**37**(2):281-339

[40] Wilson SS, Wiens ME, Smith JG. Antiviral mechanisms of human defensins. Journal of Molecular Biology. 2013;**425**(24):4965-4980

[41] Wiegand I, Hilpert K, Hancock REW. Agar and broth dilution methods to determine the minimal inhibitory concentration (MIC) of antimicrobial substances. Nature Protocols. 2008;**3**(2):163

[42] Wu X, Wang Z, Li X, Fan Y, He G, Wan Y, et al. In vitro and in vivo activity of antimicrobial peptides developed using an amino acid-based activity prediction method. Antimicrobial Agents and Chemotherapy. 2014. AAC--02823

[43] Giacometti A, Cirioni O, Barchiesi F, Del Prete MS, Fortuna M, Caselli F, et al. In vitro susceptibility tests for cationic peptides: Comparison of broth microdilution methods for bacteria that grow aerobically. Antimicrobial Agents and Chemotherapy. 2000;**44**(6):1694-1696

[44] Gopal R, Park JS, Seo CH, Park Y. Applications of circular dichroism for structural analysis of gelatin and antimicrobial peptides. International Journal of Molecular Sciences. 2012;**13**(3):3229-3244

[45] Porcelli F, Ramamoorthy A, Barany G, Veglia G. On the role of NMR spectroscopy for characterization of antimicrobial peptides. In: Membrane Proteins. Springer; 2013. pp. 159-180

[46] Jeong J-H, Kim J-S, Choi S-S, Kim Y. NMR structural studies of antimicrobial peptides: LPcin analogs. Biophysical Journal. 2016;**110**(2):423-430

[47] Kallick DA, Tessmer MR, Watts CR, Li C-Y. The use of dodecylphosphocholine micelles in solution NMR. Journal of Magnetic Resonance: Series B. 1995;**109**(1):60-65

[48] De Angelis AA, Grant CV, Baxter MK, McGavin JA, Opella SJ, Cotten ML. Amphipathic antimicrobial piscidin in magnetically aligned lipid bilayers. Biophysical Journal. 2011;**101**(5):1086-1094

[49] Carr JK, Wang L, Roy S, Skinner JL. Theoretical sum frequency generation spectroscopy of peptides. The Journal of Physical Chemistry B. 2014;**119**(29):8969-8983

[50] Golbek TW, Franz J, Elliott Fowler J, Schilke KF, Weidner T, Baio JE. Identifying the selectivity of antimicrobial peptides to cell membranes by sum frequency generation spectroscopy. Biointerphases. 2017;**12**(2):02D406

[51] Wakabayashi N, Yano Y, Kawano K, Matsuzaki K. A pH-dependent charge reversal peptide for cancer targeting. European Biophysics Journal. 2017;**46**(2):121-127

[52] Huang Y, Huang J, Chen Y. Alpha-helical cationic antimicrobial peptides: Relationships of structure and function. Protein & Cell. 2010;**1**(2):143-152

[53] Rehaume LM, Hancock REW. Neutrophil-derived defensins as modulators of innate immune function. Critical Reviews in Immunology. 2008;**28**(3)

[54] Lehrer RI, Ganz T. Defensins of vertebrate animals. Current Opinion in Immunology. 2002;**14**(1):96-102

[55] Shai Y. Mode of action of membrane active antimicrobial peptides. Biopolymers. 2002;**66**(4):236-248

[56] Hill CP, Yee J, Selsted ME, Eisenberg D. Crystal structure of defensin HNP-3, an amphiphilic dimer: Mechanisms of membrane permeabilization. Science (80-). 1991;**251**(5000):1481-1485

[57] Wilson CL, Ouellette AJ, Satchell DP, Ayabe T, López-Boado YS, Stratman JL, et al. Regulation of intestinal α-defensin activation by the metalloproteinase matrilysin in innate host defense. Science (80-). 1999;**286**(5437):113-117

[58] Valore EV, Ganz T. Posttranslational processing of defensins in immature human myeloid cells. Blood. 1992;**79**(6):1538-1544

[59] Kagan BL, Selsted ME, Ganz T, Lehrer RI. Antimicrobial defensin peptides form voltage-dependent ion-permeable channels in planar lipid bilayer membranes. Proceedings of the National Academy of Sciences. 1990;**87**(1):210-214

[60] Selsted ME, Ouellette AJ. Mammalian defensins in the antimicrobial immune response. Nature Immunology. 2005;**6**(6):551

[61] Lehrer RI, Ganz T, Szklarek D, Selsted ME. Modulation of the in vitro candidacidal activity of human neutrophil defensins by target cell metabolism and divalent cations. The Journal of Clinical Investigation. 1988;**81**(6):1829-1835

[62] Zeya HI, Spitznagel JK. Antibacterial and enzymic basic proteins from leukocyte lysosomes: separation and identification. Science (80-). 1963;**142**(3595):1085-1087

[63] Anunthawan T, de la Fuente-Núñez C, Hancock REW, Klaynongsruang S. Cationic amphipathic peptides KT2 and RT2 are taken up into bacterial cells and kill planktonic and biofilm bacteria. Biochimica et Biophysica Acta (BBA): Biomembranes. 2015;**1848**(6):1352-1358

[64] Malanovic N, Lohner K. Gram-positive bacterial cell envelopes: The impact on the activity of antimicrobial peptides. Biochimica et Biophysica Acta (BBA): Biomembranes. 2016;**1858**(5):936-946

[65] Weinberg A, Krisanaprakornkit S, Dale BA. Epithelial antimicrobial peptides: Review and significance for oral applications. Critical Reviews in Oral Biology and Medicine. 1998;**9**(4):399-414

[66] Daher KA, Selsted ME, Lehrer RI. Direct inactivation of viruses by human granulocyte defensins. Journal of Virology. 1986;**60**(3):1068-1074

[67] Zhang H, Porro G, Orzech N, Mullen B, Liu M, Slutsky AS. Neutrophil defensins mediate acute inflammatory response and lung dysfunction in dose-related fashion. American Journal of Physiology-Cell Physiology. 2001;**280**(5):L947-L954

[68] Wimley WC, Selsted ME, White SH. Interactions between human defensins and lipid bilayers: Evidence for formation of multimeric pores. Protein Science. 1994;**3**(9):1362-1373

[69] Hoover DM, Rajashankar KR, Blumenthal R, Puri A, Oppenheim JJ, Chertov O, et al. The structure of human β-defensin-2 shows evidence of higher order oligomerization.

The Journal of Biological Chemistry. 2000;**275**(42):32911-32918

[70] Aarbiou J, Verhoosel RM, van Wetering S, de Boer WI, van Krieken JHJM, Litvinov SV, et al. Neutrophil defensins enhance lung epithelial wound closure and mucin gene expression in vitro. American Journal of Respiratory Cell and Molecular Biology. 2004;**30**(2):193-201

[71] Huang LC, Redfern RL, Narayanan S, Reins RY, McDermott AM. In vitro activity of human β-defensin 2 against *Pseudomonas aeruginosa* in the presence of tear fluid. Antimicrobial Agents and Chemotherapy. 2007;**51**(11):3853-3860

[72] Yang D, Chen Q, Chertov O, Oppenheim JJ. Human neutrophil defensins selectively chemoattract naive T and immature dendritic cells. Journal of Leukocyte Biology. 2000;**68**(1):9-14

[73] Barnes L, Cooper I. Biomaterials and Medical Device-Associated Infections. Elsevier; 2014

[74] Hetrick EM, Schoenfisch MH. Reducing Implant-Related Infections: Active Release Strategies. Chemical Society Reviews. 2006;**35**(9):780-789

[75] Weinstein RA, Darouiche RO. Device-associated infections: A macroproblem that starts with microadherence. Clinical Infectious Diseases. 2001;**33**(9):1567-1572

[76] Percival SL, Suleman L, Vuotto C, Donelli G. Healthcare-associated infections, medical devices and biofilms: Risk, tolerance and control. Journal of Medical Microbiology. 2015;**64**(4):323-334

[77] Zhang L, Keogh S, Rickard CM. Reducing the risk of infection associated with vascular access devices through nanotechnology: A perspective. International Journal of Nanomedicine. 2013;**8**:4453

[78] Maillard J-Y. Antimicrobial biocides in the healthcare environment: Efficacy, usage, policies, and perceived problems. Therapeutics and Clinical Risk Management. 2005;**1**(4):307

[79] Yin J, Luan S. Opportunities and challenges for the development of polymer-based biomaterials and medical devices. Regenerative Biomaterials. 2016;**3**(2):129-135

[80] Banerjee I, Pangule RC, Kane RS. Antifouling coatings: Recent developments in the design of surfaces that prevent fouling by proteins, bacteria, and marine organisms. Advanced Materials. 2011;**23**(6):690-718

[81] Bruellhoff K, Fiedler J, Möller M, Groll J, Brenner RE. Surface coating strategies to prevent biofilm formation on implant surfaces. The International Journal of Artificial Organs. 2010;**33**(9):646-653

[82] Matsuzaki K. Control of cell selectivity of antimicrobial peptides. Biochimica et Biophysica Acta (BBA): Biomembranes. 2009;**1788**(8):1687-1692

[83] Deslouches B, Phadke SM, Lazarevic V, Cascio M, Islam K, Montelaro RC, et al. De novo generation of cationic antimicrobial peptides: Influence of length and tryptophan substitution on antimicrobial activity. Antimicrobial Agents and Chemotherapy. 2005;**49**(1):316-322

[84] Eckert R. Road to clinical efficacy: Challenges and novel strategies for antimicrobial peptide development. Future Microbiology. 2011;**6**(6):635-651

[85] Werle M, Bernkop-Schnürch A. Strategies to improve plasma half life time of peptide and protein drugs. Amino Acids. 2006;**30**(4):351-367

[86] Marr AK, Gooderham WJ, Hancock REW. Antibacterial peptides for therapeutic use: Obstacles and realistic outlook. Current Opinion in Pharmacology. 2006;**6**(5):468-472

[87] Seebach D, Beck AK, Bierbaum DJ. The world of β-and γ-peptides comprised of homologated proteinogenic amino acids and other components. Chemistry & Biodiversity. 2004;**1**(8):1111-1239

[88] Tugyi R, Uray K, Iván D, Fellinger E, Perkins A, Hudecz F. Partial D-amino acid substitution: Improved enzymatic stability and preserved Ab recognition of a MUC2 epitope peptide. Proceedings of the National Academy of Sciences. 2005;**102**(2):413-418

[89] Gentilucci L, De Marco R, Cerisoli L. Chemical modifications designed to improve peptide stability: Incorporation of non-natural amino acids, pseudo-peptide bonds, and cyclization. Current Pharmaceutical Design. 2010;**16**(28):3185-3203

[90] Bagheri M, Beyermann M, Dathe M. Immobilization reduces the activity of surface-bound cationic antimicrobial peptides with no influence upon the activity spectrum. Antimicrobial Agents and Chemotherapy. 2009;**53**(3):1132-1141

[91] Chen R, Cole N, Willcox MDP, Park J, Rasul R, Carter E, et al. Synthesis, characterization and in vitro activity of a surface-attached antimicrobial cationic peptide. Biofouling. 2009;**25**(6):517-524

[92] Boulmedais F, Frisch B, Etienne O, Lavalle P, Picart C, Ogier J, et al. Polyelectrolyte multilayer films with pegylated polypeptides as a new type of anti-microbial protection for biomaterials. Biomaterials. 2004;**25**(11):2003-2011

[93] Shukla A, Fleming KE, Chuang HF, Chau TM, Loose CR, Stephanopoulos GN, et al. Controlling the release of peptide antimicrobial agents from surfaces. Biomaterials. 2010;**31**(8):2348-2357

[94] Ferreira L, Zumbuehl A. Non-leaching surfaces capable of killing microorganisms on contact. Journal of Materials Chemistry. 2009;**19**(42):7796-7806

[95] Xiao M, Jasensky J, Gerszberg J, Chen J, Tian J, Lin T, et al. Chemically immobilized antimicrobial peptide on polymer and self-assembled monolayer substrates. Langmuir. 2018;**34**(43):12889-12896

[96] Glinel K, Jonas AM, Jouenne T, Leprince J, Galas L, Huck WTS. Antibacterial and antifouling polymer brushes incorporating antimicrobial peptide. Bioconjugate Chemistry. 2008;**20**(1):71-77

[97] Gabriel M, Nazmi K, Veerman EC, Nieuw Amerongen AV, Zentner A. Preparation of LL-37-grafted titanium surfaces with bactericidal activity. Bioconjugate Chemistry. 2006;**17**(2):548-550

[98] Onaizi SA, Leong SSJ. Tethering antimicrobial peptides: Current status and potential challenges. Biotechnology Advances. 2011;**29**(1):67-74

[99] Appendini P, Hotchkiss JH. Surface modification of poly (styrene) by the attachment of an antimicrobial peptide. Journal of Applied Polymer Science. 2001;**81**(3):609-616

[100] Hilpert K, Elliott M, Jenssen H, Kindrachuk J, Fjell CD, Körner J, et al. Screening and characterization of surface-tethered cationic peptides for antimicrobial activity. Chemistry & Biology. 2009;**16**(1):58-69

[101] Spindler EC, Hale JDF, Giddings TH, Hancock REW, Gill RT. Deciphering the mode of action

of a synthetic antimicrobial peptide (Bac8c). Antimicrobial Agents and Chemotherapy. 2011

[102] Li Y, Wei S, Wu J, Jasensky J, Xi C, Li H, et al. Effects of peptide immobilization sites on the structure and activity of surface-tethered antimicrobial peptides. Journal of Physical Chemistry C. 2015;**119**(13):7146-7155

[103] Soares JW, Kirby R, Doherty LA, Meehan A, Arcidiacono S. Immobilization and orientation-dependent activity of a naturally occurring antimicrobial peptide. Journal of Peptide Science. 2015;**21**(8):669-679

[104] Xiao M, Jasensky J, Foster L, Kuroda K, Chen Z. Monitoring antimicrobial mechanisms of surface-immobilized peptides in situ. Langmuir. 2018;**34**(5):2057-2062

[105] Han X, Zheng J, Lin F, Kuroda K, Chen Z. Interactions between surface-immobilized antimicrobial peptides and model bacterial cell membranes. Langmuir. 2017;**34**(1):512-520

[106] Costa F, Carvalho IF, Montelaro RC, Gomes P, Martins MCL. Covalent immobilization of antimicrobial peptides (AMPs) onto biomaterial surfaces. Acta Biomaterialia. 2011;**7**(4):1431-1440

[107] Holmberg K, Bergström K, Stark M-B. Immobilization of proteins via PEG chains. In: Poly (Ethylene Glycol) Chemistry. Springer; 1992. pp. 303-324

[108] Cho W-M, Joshi BP, Cho H, Lee K-H. Design and synthesis of novel antibacterial peptide-resin conjugates. Bioorganic & Medicinal Chemistry Letters. 2007;**17**(21):5772-5776

[109] Veronese FM, Pasut G. PEGylation, successful approach to drug delivery. Drug Discovery Today. 2005;**10**(21):1451-1458

[110] Guiotto A, Pozzobon M, Canevari M, Manganelli R, Scarin M, Veronese FM. PEGylation of the antimicrobial peptide nisin A: Problems and perspectives. Farmácia. 2003;**58**(1):45-50

Chapter 5

Determination of Substrate Specificity of the Purified Novel Plant Cysteine Protease Solanain From the Latex of *Vallaris solanacea*

Silpa Somavarapu

Abstract

Peptide synthesis refers to the production of peptides. Proteases are the enzymes that degrade proteins. They hold first place in the world market of enzymes, estimated at ~US$3 billion. Proteases are distributed widely in different parts of the biological sources. In occurrence of proteases, plant kingdom occupies the highest rank. The increasing demand for proteases and the need for economical production of commercially useful industrial proteases from novel sources are taken into consideration. A novel plant latex cysteine protease namely *Vallaris solanacea* was identified in Biodiversity Park, Visakhapatnam which showed maximum protease activity. It has been shown that the latex of *Vallaris solanacea* contains a high protease activity. The cysteine protease solanain was purified by fractionation with ammonium sulphate followed by DEAE-cellulose ion exchange and gel chromatography. Specificity studies towards synthetic peptide and ester substrates by the protease purified from the latex of *Vallaris solanacea* were performed. The purified solanain exhibited broad specificity similar to other cysteine proteases. However considerable differences were also noticed in the rate of hydrolysis and specificity towards simple peptide substrates.

Keywords: peptide synthesis, proteases, *Vallaris solanacea*, specificity studies, cysteine proteases, peptide substrates

1. Peptide synthesis

Peptide synthesis refers to the production of peptides. Peptides and proteins are linear polymers of amino acids linked by amide peptide bonds. Amino acids are primary amines that contain an anomeric carbon that is connected to a hydrogen atom (H), an amino (NH_3) group, a carboxyl group (COOH) and a variable side group (R) (**Figure 1**).

Peptides are chemically synthesized by the condensation reaction of the carboxyl group of one amino acid to the amino group of another. The formation of peptide bonds by linking amino acids together dates back to 100 years. The first peptides to be synthesized, including oxytocin and insulin took about 50–60 years

IntechOpen

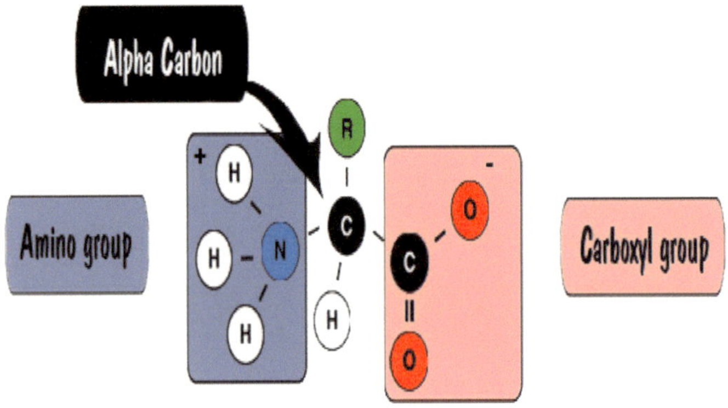

Figure 1.
Peptide structure.

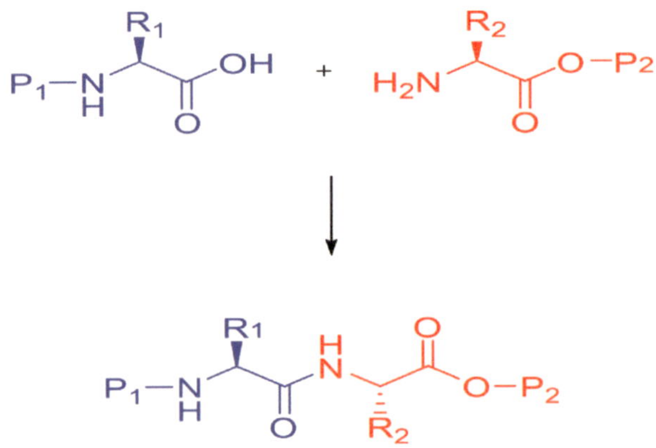

Figure 2.
Coupling of two amino acids in solution. The unprotected amine of one reacts with the unprotected carboxylic acid group of the other to form a peptide bond.

which clearly indicates chemical synthesis of peptides is a difficult task [1]. But advances in protein synthesis today made peptide synthesis easier today which has varied applications in high-throughput biological research and drug development [2] (**Figure 2**).

2. Peptide synthesis

They are two methods of peptide synthesis, namely

- Liquid-phase peptide synthesis
- Solid-phase peptide synthesis

Liquid-phase peptide synthesis is the classical method that is slow and labour intensive in which separate peptides are synthesized and then coupled together to create larger peptides. In this method chemical group is used to protect the C-terminus of the first amino acid.

Solid-phase peptide synthesis is the most common method where the C-terminus of the first amino acid is coupled to an activated solid support, such as polystyrene or polyacrylamide. This method is advantageous as the resin acts as the

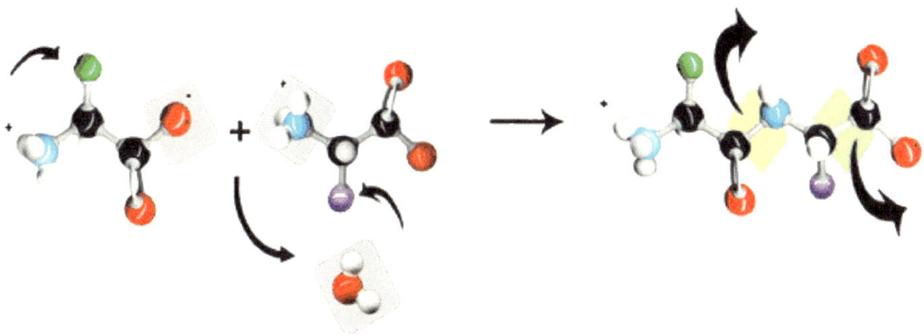

Figure 3.
Peptide bonds created with peptide coupling agents.

C-terminal protecting group and it is a rapid method to separate the growing peptide product from the different reaction mixtures during synthesis.

These two ways are combined in an exceedingly method known as native chemical tying. LifeTein's customary amide synthesis method uses the solid part. The liquid-phase approach is employed for the synthesis of short peptides, like di- and tripeptides, and C-terminally changed peptides, like accelerator substrates (**Figure 3**).

3. Solid phase synthesis

The important method for the synthesis of peptides in the lab is known as solid-phase peptide synthesis (SPPS) [3]. Merrifield [4] developed this method of SPPS which allows the rapid assembly of a peptide chain through successive reactions of amino acid derivatives on an insoluble porous support. Unlike ribosome protein synthesis, artificial synthesis builds peptides in the C to N direction. During solid-phase peptide synthesis, each peptide is anchored to an insoluble polymer at the C-terminus.

3.1 Solid support

Solid support in SPPS is Polystyrene, a styrene cross-linked with 1–2% divinylbenzene which is a popular carrier resin in SPPS. Other common gel-type supports include polyacrylamide and polyethylene glycol (PEG). Polystyrene is chemically inert under SPSS conditions. The solid support helps in anchoring the amino acids and formation of peptide.

3.2 Protecting groups

Amino acids contain side chains with different functional groups with different reactivity. Thus different protecting groups are required. A major problem in peptide synthesis is the side reactions due to multiple reactive groups in amino acids. In order to perform peptide formation with minimal side reactions or to protect the functional groups from non-specific reactions reactive groups in the amino acids need to be blocked or protected. For this many chemical groups have been developed that bind to the amino acids and protect it. They are two types of protecting groups, namely

- Temporary protecting groups
- Permanent protecting groups

3.2.1 Temporary protecting groups

N-terminal ends of amino acids are protected by groups called 'temporary' protecting groups as they are easily removed to allow peptide bond formation.

3.2.2 Permanent protecting groups

But Side chain protecting groups are known as permanent protecting groups as they remain at all the multiple cycles of chemical treatment during the synthesis phase and they can be removed only on treatment with strong acids after synthesis is complete (**Figure 4**).

Peptide synthesis involves coupling of carboxyl group and the incoming amino acid to the N-terminus of the growing peptide chain. It is a step wise method and also a cyclic process. It involves the following steps.

1. Blocking of amino group of amino acid by N-terminal protecting groups

Prior to protein synthesis, individual amino acids are reacted with these protecting groups. Two common N-terminal protecting groups are tert-butoxycarbonyl (Boc) and 9-fluorenylmethoxycarbonyl (Fmoc), and each group has distinct characteristics that determine their use (**Table 1**; **Figure 5**).

2. Blocking of carbonyl group of amino acid by C-terminal protecting groups

Similarly the carbonyl group of amino acid is protected by C-terminal protecting group. The C-terminal protecting group depends on the type of peptide

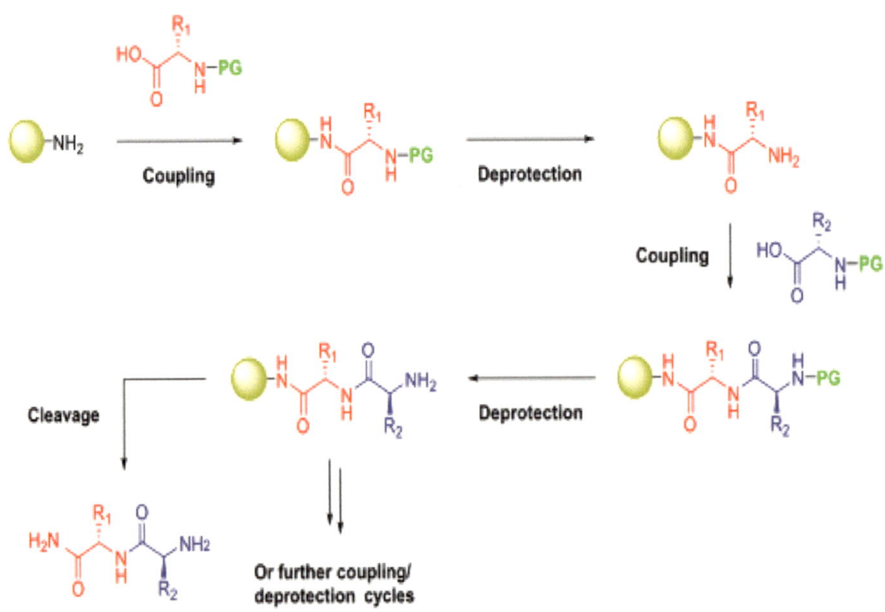

Figure 4.
Solid-phase synthesis of a dipeptide victimization associate degree (amine-functionalized) organic compound. The N-terminal protective cluster (PG) will be Fmoc or Boc, reckoning on the protective cluster theme used (see below). The amino alkanoic acid facet chains (R1, R2, etc.) are orthogonally protected (not shown).

S. No	Fmoc	Boc
1	Routine synthesis	Requires special equipment
2	Relatively safe	Potentially dangerous
3	Acid-sensitive peptides and derivatives	Base-labile peptides
4	Frequent aggregation	Moderate aggregation
5	TFA final deprotection	HF final deprotection

Table 1.
Difference between Fmoc and Boc.

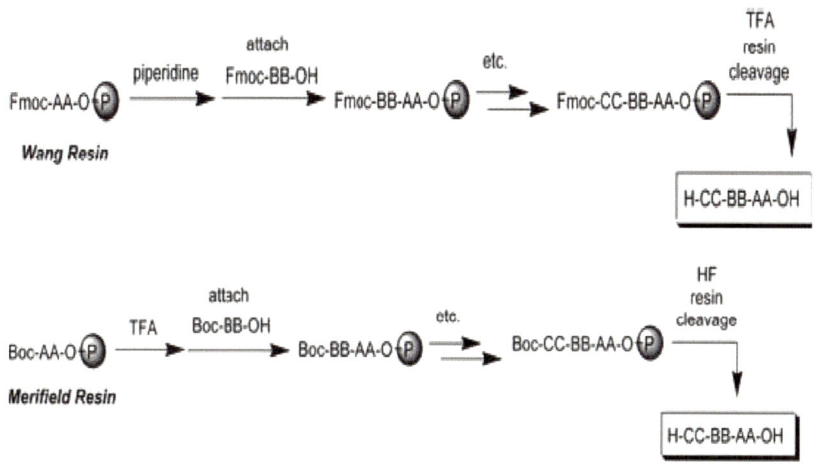

Figure 5.
Fmoc strategy (Wang resin) and Boc strategy (Merrifield resin).

synthesis used. For liquid-phase peptide synthesis C-terminus of the first amino acid is protected by chemical reagent while in solid-phase peptide synthesis solid support (resin) acts as the protecting group for the C-terminal amino acid (**Figure 6**).

3. Coupling of the protected amino acids to form a peptide

Once peptide synthesis is completed, all the protecting groups are removed from the nascent peptides. First C-terminal carboxylic acid is activated using carbodiimides like dicyclohexylcarbodiimide (DCC) or diisopropylcarbodiimide (DIC). These coupling reagents respond with the carboxyl gathering to shape an exceptionally receptive O-acylisourea moderate that is immediately dislodged by nucleophilic assault from the deprotected essential amino gathering on the

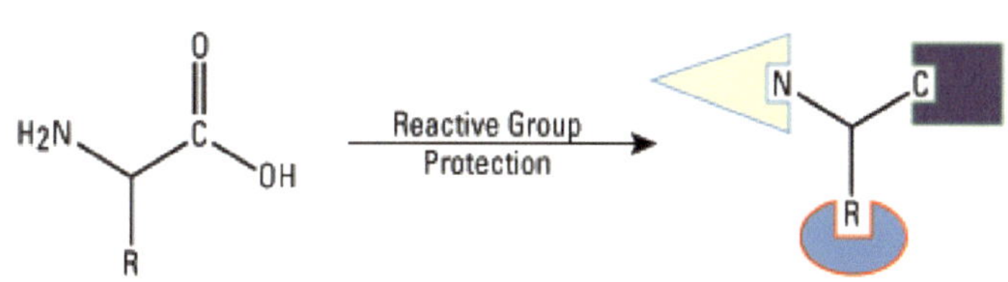

Figure 6.
Amino acid functional group protection.

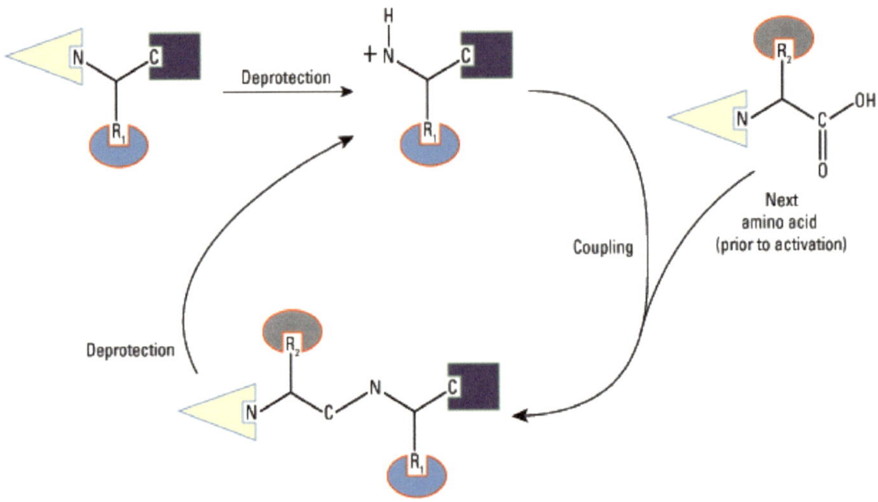

Figure 7.
Diagram of peptide synthesis. Peptide bond formation between the deprotected N-terminus of the first amino acid and the activated C-terminus of the incoming amino acid. This cycle of deprotection and coupling is repeated until the full-length peptide is formed.

N-end of the developing peptide chain to frame the beginning peptide security. To affirm total coupling a test is performed called Kaiser Test (**Figure** 7).

4. Deblocking the amino group of amino acid

Removal of specific protecting groups from the newly added amino acid immediately after coupling to allow the next incoming amino acid to bind to the growing peptide chain is called deprotection. Boc is removed using moderately strong acid such as trifluoroacetic acid (TFA) while Fmoc is a base-labile protecting group that is removed with a mild base such as piperidine.

4. Deblocking the carbonyl group of amino acid

The synthetic peptide purification is by compounds like water, anisol or thiol derivatives are added in excess during the deprotection step to react with any of

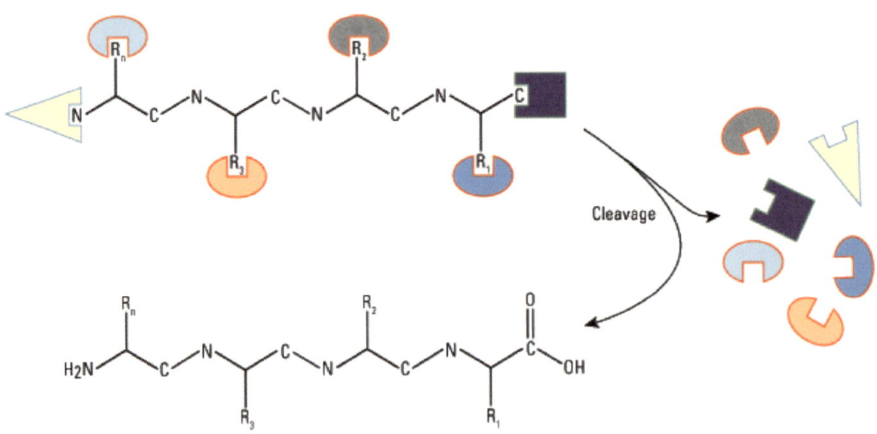

Figure 8.
Diagram of peptide cleavage after synthesis. The remaining N-terminal protecting groups, all side-chain protecting groups and the C-terminal protecting group or solid support are removed by strong acid treatment after peptide synthesis is completed.

these free reactive species. This cycle of deprotection and coupling is repeated. The remaining protecting groups are cleaved by acidolysis, using; strong acids such as hydrogen fluoride (HF), hydrogen bromide (HBr) or trifluoromethane sulphonic acid (TFMSA) (**Figure 8**).

5. Peptide purification

Peptides are purified by the following techniques namely Ion exchange chromatography (IEC), Partition chromatography, Size-exclusion chromatography, Reverse-phase chromatography (RPC) and High-performance liquid chromatography (HPLC).

6. Applications of synthetic peptides

The Peptide synthesis (**Figure 9**) holds varied applications including

- Synthetic peptides are used to study enzyme-substrate interactions within important enzyme classes such as kinases and proteases, which play a crucial role in cell signalling.
- The development of epitope-specific antibodies against pathogenic proteins.
- The study of protein functions and the identification and characterization of proteins.

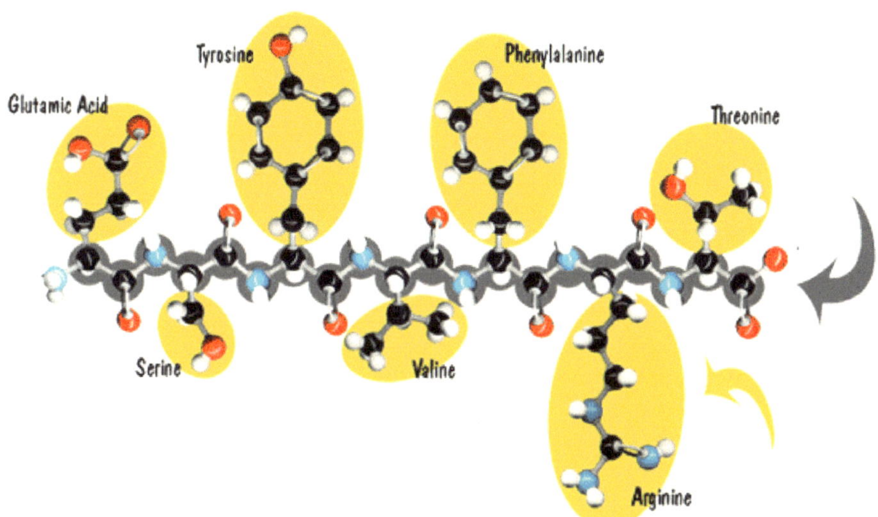

Figure 9.
Polypeptide formation.

7. Proteases

For about a century, the enzymes that play the central role in the degradation of proteins by hydrolysing peptide bonds have been known as 'proteases' and the term protease is therefore equivalent to 'peptide hydrolase'. They hold first place in the world market of enzymes, estimated at ~US$3 billion [5]. Proteases are distributed widely in different parts of the biological sources. In occurrence of proteases, plant kingdom occupies the highest rank (43.85%) followed by bacteria (18.09%), fungi (15.08%), animals (11.15%), algae (7.42%) and viruses (4.41%). According to

Family, species	Name, EC-number	Reference(s)
In latices	Papain, 3.4.22.2	[9, 11, 19]
(I) Caricaceae	Chymopapain,	Kunimitsu and Yasunobu, 1970, Robinson, 1975;
1. *Carica papaya*	3.4.22.6	Khan and Polgar, 1983
2. *Pileus mexicanus*	Papayapeptidase-I	Lynn, 1979; Lynn and Yaguchi, 1979; Polgar, 1981;
(II) Moraceae	(A and B)	Baines and Brocklechurt, 1982
3. *Ficus carica*	Mexicain, 3.4.99.14	Castaneda-Agullo et al., 1942; Soriano et al., 1975
4. *Ficus racemosa*	Ficin, 3.4.22.3	Sgarbieri et al., 1964; Kramer and Whitaker, 1964;
(III) Asclepiadaceae	Protease	Sugiura and Sasaki, 1974
5. *Asclepias speciosa*	Asclepain, 3.4.22.7	Devaraj et al., 2008
6. *Funastrum clausum*	Funastrain c II	Greenberg and Winnick, 1940
(IV) Apocynaceae	Tabernamontanain,	Morcelle et al., 2004
7. *Tabernaemontana*	3.4.99.23	Jaffe, 1943
grandiflora	Indicain	Singh, 2008
(V) Urticaceae	Milin	Yadav et al., 2006
8. *Morus indica*	Taraxilisin	Rudenskaya et al., 1998
(VI) Euphorbiaceae	Fruit bromelain,	Cooreman et al., 1976; Murachi 1970; 1976
9. *Euphorbia milii*	3.4.22.4	Messing and Santoro, 1960
(VII) Asteraceae	Pinguinain,	Agundis et al., 1977
10. *Taraxacum*	3.4.99.18	Arcus, 1959; McDowall, 1973; Brocklehurst et al., 1981
officinale	Hemisphaericin	Murachi, 1970; 1976; Heinicke and Gortner, 1957
In fruits	Actindin, 3.4.22.14	[12]
(IX) Bromeliaceae	Stem bromelain,	Daley and Vines, 1978
30. *Ananas comosus*	3.4.22.4	Yamaguchi et al., 1982
31. *Bromelia penguin*	Ananain	Yamaguchi et al., 1982
32. *Bromelia hemisph*	Leaf proteases	
(X) Actinidiaceae	… … … …	
33. *Actinidia chinensis*	… … … …	
In vegetative organs		
(XI) Bromeliaceae		
34. *Ananas comosus*		
(XII) Zingiberaceae		
35. *Zingiber officinale*		
36. Miut		

Table 2.
List of plant cysteine proteases and their sources.

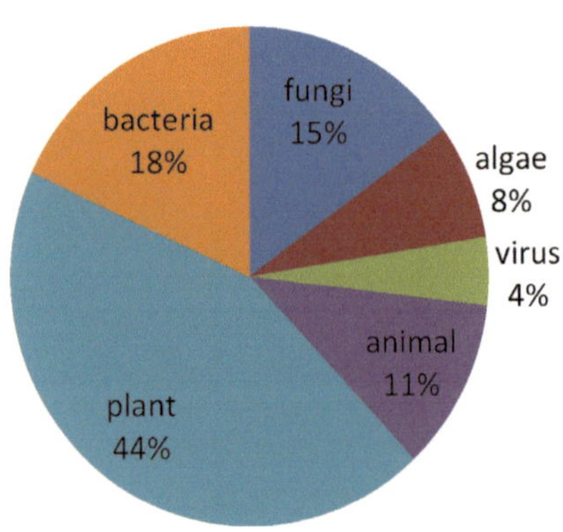

Figure 10.
Distribution of proteases.

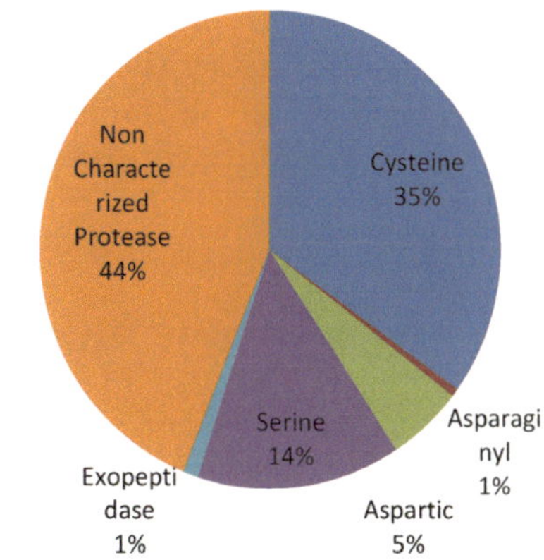

Figure 11.
Plant proteases.

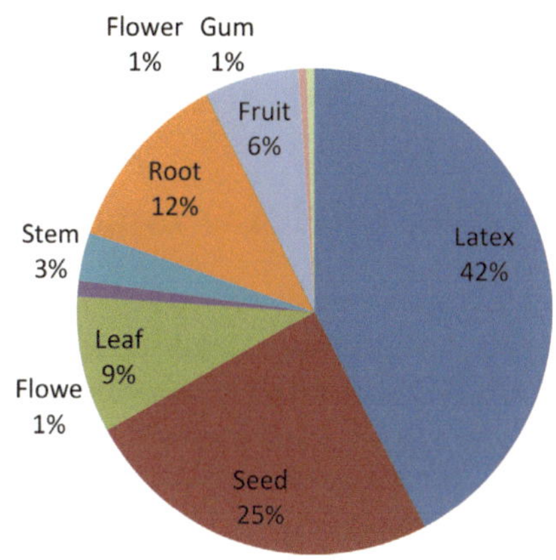

Figure 12.
Distribution of proteases in plant parts.

Actinidin	Asp-Ile-Lys-Ser-Gln-Gly-Glu-Cys-Gly-Gly-Cys-Trp
Papain	Pro-Val-Lys-Asn-Gln-Gly-Ser-Cys-Gly-Ser-Cys-Trp
Ficin	Pro-Ile-Arg-Gln-Gln-Gly-Gln-Cys- Gly-Ser-Cys-Trp
Stem-bromelain	Asn-Gln-Asp-Pro-Cys-Gly-Ala-Cys-Trp
Fruit –bromelain	Asn-Gln-Asn-Pro-Cys-Gly-Ala-Cys
Chymopapain	Lys-Arg-Val-Pro-Asp-Ser-Gly-Glu-Cys-Tyr

Papain	Val-Gly-Pro-Cys-Gly-Asn-Lys-Val-His-Ala-Val-Ala-Ala-Val-Gly-Tyr
Ficin	Thr-Pro-Cys-Gly-Thr-Ser-Leu-Asp-His-Ala-Val-Ala-Leu
Stem- bromelain	His-Ala-Val-Thr-Ala-Ile-Gly-Tyr
Actinidin	Pro-Cys-Gly-Thr-Ala-Val-Asp-His-Ala-Ile-Val-Ile-Val-Gly

Figure 13.
Amino acid sequence around the active site cysteine and histidine residues (in bold) of some plant cysteine proteases.

Barrett and McDonald [6], plant proteases are classified into serine, cysteine, aspartic and metalloproteases. Cysteine proteases (EC 3.4.2.2) are found in bacteria [7], eukaryotic microorganisms [8], plants [9] and animals. Cysteine proteases are represented by 70 families belonging to 12 different classes [10] (**Table 2**; **Figures 10–13**).

8. Screening of plant lattices for novel plant latex cysteine protease

Taking into consideration, the increasing demand for proteases and the need for economical production of commercially useful industrial proteases from novel sources. A number of plant lattices belonging to different plant families have been collected in and around Visakhapatnam and screened for protease activity. Their activities were assayed immediately and were stored in ice for further investigations. A novel plant latex cysteine protease namely *Vallaris solanacea* was identified in Biodiversity Park, Visakhapatnam which showed maximum protease activity. Preliminary studies on protease activity from the latex of *Vallaris solanacea* were carried out. The protease was purified and characterized. Specificity studies towards synthetic peptide and ester substrates by the protease purified solanain.

9. Substrate specificity of the purified solanain from the latex of *Vallaris solanacea*

Plant lattices are rich source of proteases. Latex of *Vallaris solanacea* also showed high protease activity. The cysteine protease solanain was purified by ammonium sulphate precipitation followed by DEAE-cellulose ion exchange and gel chromatography. The purpose of the present study is determination of specificity of purified solanain towards synthetic peptide and ester substrates. Extensive and more systematic studies [13–16] have been made on papain, ficin [17] and bromelain [18].

10. Materials and methods

10.1 Preparation of peptide solutions

First step is preparation of 10 mM peptide solution having tryptophan, tyrosine and phenylalanine. Peptides were solubilized using dilute acid, the assay mixture contained higher buffer concentration viz. 0.15 M.

10.1.1 Solanain action on peptides

Peptide was incubated by solanain in 0.05 M Phosphate buffer (pH 7.0), 10 mm β-mercaptoethanol, 2 mM EDTA, 2.5 mm amide and 0.5 mg to 1.0 mg/ml catalyst protein. Incubation was administered for 48 h at 30°C. The reaction was detected by sampling the digestible mixture on aid. For every digestion mixture, a bearing while not substrate or without catalyst was analysed. Skinny layer activity was performed as delineated by Renderath (1963).

10.1.2 Preparation of colloid silica gel G plates

About 25 weight unit of colloid Silica Gel G was mixed with 50 cm^3 of H_2O and stirred smartly during a closed round shape flask till completely distributed. The suspension was transferred into Stahl's mechanical spreader adjusted to 250 μ thickness, victimization the spreader; layers were ready on 20 × 20 cm glass plates. The gel was allowed to dry for a couple of minutes and so activated by drying in associate degree kitchen appliance at 1100°C for half-hour.

10.1.3 Developing the chromatographic plate

About 5–10 μl of the sample was loaded. Solvent used was 4:1:1 butanol: carboxylic acid: water (v/v/v). After development the plates were removed, dried and detected by spraying with 0.2% ninhydrin in butanol: ethanoic acid (95, 5 v/v) mixture. Rf values of the spots were calculated. Amino acids were identified.

11. Results and discussion

Studies on substrate specificity were done and results were tabulated. Solanain was capable of hydrolysing peptide bonds involving the amino groups of hydrophilic amino acid residues (peptides 1 to 5) and incapable of peptide bonds involving the groups of deliquescent organic compound residues (peptides one to 5) and incapable of hydrolysing amide bonds wherever amino group was given by a

S. No	Peptides	Protease activity
Simple dipeptides		
1	Gly – Gly	++
2	Gly – L - α- Ala	+
3	Gly – L – Asn	+++
4	Gly – DL – Asn	+
5	Gly – D – Asn	+++
6	Gly – L – Leu	–
7	Gly – L - β- phe	–
8	Gly – L – Trp	–
9	L – Ala – L – Met	–
10	L – Leu– L – Met	+++
11	L – Trp – Gly	–
12	L – Trp – L – Tyr	–
13	L – Tyr – Gly	–
N-aryl-dipeptides		
14	N – Z – L – Glu – L – phe	+++
15	N – Z – L – Glu – L – Tyr	++
16	N – Z – L – Ileu – L – Met	–
17	N – Z – L – Met – Gly – OEt*	++++
18	Hippuryl – L – Arg	++
Tripeptides		
19	Gly – Gly – Gly	–
20	L-Glu-L-Val-L-Phe	–
21	L – Leu- Gly – Gly	+

**Both ester bond and the peptide bond of N – Z – L – Met – Gly – OEt were hydrolysed.*

Table 3.
Reaction of di- and tripeptides by refined Solanain.

Substrate		$10^8(E_O)(M)$	V_O/E_O Moles/sec/mole enzyme	% Relative activity*
Amino acid side chain	**p-Nitrophenyl ester**			
Nil	Z-Gly-Onp	1.63	5.29	10
Unbranched	Z-L-Ala-Onp	3.27	16.94	320.00
Branched	Z-L-Val-ONP	16.30	0.48	9.07
	Z-L-Leu-Onp	3.27	2.14	10.45
	Z-L-Ileu-Onp	16.30	0.21	3.97
Aromatic	Z-L-phe-Onp	3.27	1.02	09.28
	Z-L-Tyr-Onp	3.27	1.69	31.94
	Z-L-Trp-Onp	3.27	0.74	13.99
Imino acid	Z-L-pro-Onp	163.00	0.0	0.0
	N,N^1-di-Z-L Lys-Onp	163.00	0.0	0.0

**Calculated taking activity toward Z-Gly-ONp as 100%.*

Table 4.
Hydrolysis of various synthetic ester substrates with purified solanain.

hydrophobic residues (peptides half dozen to 9) with a large aspect chain e.g., leucine, essential amino acid, essential amino acid and essential amino acid. However, if the organic compound contributory the group of the bond was aromatic the bond was not hydrolysed although the C-terminal amino acid was deliquescent. Neither L-Trp-Gly nor L-Tyr-Gly was hydrolysed. It seems that the presence of a deliquescent organic compound at the N-terminal finish is not needed for the protein.

In **Table 3**, the results of the speed measurements given for 10 parts as V_o/E_o wherever V_o is initial rate and E_o is concentration of the protein in moles. The concentration of substrate used was one.56 × 10-4 M. If the metric linear unit values of the p-nitrophenyl esters of CBZ-amino acids were of constant magnitude as for different thiol proteases like papain [9], ficin [17], bromelain [18] and ananain [12] then V_o/E_o values obtained for Glycine and amino acid were love the Kcat values of the corresponding esters of different proteases.

The refined solanain of *Vallaris* family Solanaceae showed close to identical specificity towards all the substrates.

Solanain, differs from the amino alkanoic acid proteases in having low amidase activity with BAPNA and conjointly did not show any esterase activity with BAEE. This finding suggests considerably low specificity for essential amino acid residues. It differs from enzyme, ficin and bromelain during this facet. By showing broad specificity, Solanain resembled enzyme [11], ficin [14, 17] and bromelain [18] that change a spread of amide bonds. All these, however, showed a preference for basic amino acids. The solanain of *Vallaris* potato family hydrolysed leucyl bonds with efficiency and during this respect resembled ficin [17] and differed from bromelain [18] (**Table 4**).

12. Conclusion

Substrate specificity studies showed solanain exhibited broad specificity. It showed peptidase activity, amidase activity. The enzyme was capable of catalysing the hydrolysis of p-nitrophenyl esters of amino acids. It exhibited difference in specificity towards simple peptide substrates.

Author details

Silpa Somavarapu
Department of Food Technology, Vikrama Simhapuri University, Nellore,
Andhra Pradesh, India

*Address all correspondence to: silpasomavarapu.1981@gmail.com

References

[1] Lloyd-Williams P et al. Chemical Approaches to the Synthesis of Peptides and Proteins. Vol. 278. Boca Raton: CRC Press; 1997

[2] Merrifield RB. Solid phase peptide synthesis. I. the synthesis of a tetrapeptide. Journal of the American Chemical Society. 1963;**85**: 2149-2154

[3] Chan WC, White PD. Fmoc Solid Phase Peptide Synthesis: A Practical Approach. Oxford, UK: OUP; 2000. ISBN 978-0-19-963724-9

[4] Mitchell AR. Bruce Merrifield and solid-phase peptide synthesis: A historical assessment. Biopolymers. 2008;**90**(3):175-184. DOI: 10.1002/bip.20925. PMID 18213693

[5] Leary D, Vierros M, Hamon G, Arico S, Monagle C. Marine genetic resources: A review of scientific and commercial interest. Marine Policy. 2009;**33**:183-194

[6] Barrett AJ, Mc Donald JK. Mammalian proteases: A glossary and bibliography. In: Endopeptidases. Vol. 1. London: Academic Press; 1980

[7] Morihara K. Comparative specificity of microbial proteinases. Advances in Enzymology. 1974;**41**: 179-243

[8] North MJ. Comparative biochemistry of the proteinases of eukaryotic microorganisms. Microbiological Reviews. 1982;**46**:308-340

[9] Glazer AN, Smith EL. The enzymes. In: Boyer PD, editor. Vol. 501. 1971

[10] Salas CE, Gomes MTR, Hernandez M, Lopes MTP. Plant cysteine proteinases: Evaluation of the pharmacology activity. Phytochemistry. 2008;**69**:2263-2269

[11] Arnon R. Methods in enzymology. In: Perlmann GE, Larand L, editors. New Yard, London: Academic Press; 1970. p. 226

[12] Rowan AD, Buttle DJ, Barrett AJ. Ananain: A novel cysteine proteinase found in pineapple stem. Archives of Biochemistry and Biophysics. 1988;**267**: 262-270

[13] Lowe G. The structure and mechanism of action of papain. Philosophical Transactions of the Royal Society of London, Series B. 1970;**257**: 237-248

[14] Liener IE, Friendenson B. Methods in enzymology. In: Perimann GE, London L, editors. Vol. 19. 1970. p. 261

[15] Schechter I, Berger A. On the size of the active site in proteases. I. Papain. Biochemical and Biophysical Research Communications. 1967;**27**:157-162

[16] Schechter I, Berger A. On the active site of proteases. III. Mapping the active site of papain; specific peptide inhibitors of papain. Biochemical and Biophysical Research Communications. 1968;**32**: 898-902

[17] Kortt AA, Hamilton S, Webb EC, Zerner B, Ficins (EC 3.4.22.3). Purification and characterization of the enzymatic components of the latex of *Ficus glabrata*. Biochemistry. 1974;**13**: 2023-2028

[18] Silverstein RM, Kezdy FJ. Characterization of the pineapple stem proteases (bromelain). Archives of Biochemistry and Biophysics. 1975;**167**: 678-686

[19] Hall PL, Anderson CD. Proflavine interactions with papain and ficin. I. Dye binding and its effects upon enzyme inactivation by N-alkylmaleimides. Biochemistry. 1974;**13**: 2082-2087

Chapter 6

Poly(L-Lactide) Bionanocomposites

Ali Nabipour Chakoli

Abstract

A variety of natural, synthetic, and biosynthetic polymers such as poly(L-lactide), polyhydroxyalkanoate, and poly(ε-caprolactone) are biocompatible and environmentally degradable. Biodegradability can therefore be engineered into polymers by the judicious addition of chemical linkages such as anhydride, ester, or amide bonds, among others. Poly(L-lactide) (PLLA) has attracted increasing attention due to the combination of its bioresorbability, biodegradability, biocompatibility, and shape memory effect. It has been widely applied to biomedical fields such as bone screws, surgical sutures, tissue engineering, and controlled drug delivery. Nevertheless, the PLLA is weaker than that of natural cortical bones in mechanical strength. Additionally, the ability of PLLA in cell attachment and bioactivity are weak due to its hydrophobic properties. In order to overcome the unsuitable properties of PLLA, various techniques have already been applied to modify the physical and mechanical properties of PLLA. The most significant method is to introduce some various kinds of fillers into PLLA matrix to provide reinforcing filler/PLLA composites, such as hydroxyapatite (HA), b-tricalcium phosphate, bioglass, silica gel, amorphous carbon, carbon nanotubes (CNTs), and so on.

Keywords: poly(L-lactide), nanocomposites, nanomaterials, tissue engineering

1. Introduction

Nowadays, nanoscience and nanotechnology have increased the scope of polymeric materials application, with the ultimate goal of dramatically enhanced performance [1, 2]. The most popular performance is to introduce nanoparticles into the polymer matrix to treat the polymer/nano-sized particles composites. The second is the fabrication of polymeric nanoscale materials [3, 4]. Both the mentioned approaches have been applied for various polymeric systems [5]. Based on the revolutionary researches, nanotechnology has been successfully applied to produce different kinds of biopolymer materials with valuable quality and high performance in various fields [6].

The tissue engineering, which is considered as a multidisciplinary field in medicine and industry, is emerging as the promising new approach in the reconstruction of imperfect or damaged body tissues [7, 8]. Also, tissue engineering is multidisciplinary field of integrating materials science, biotechnology, industrial engineering, and medical engineering [9]. This chapter focuses on the development of biotechnical substitutes for restoration, replacement, maintaining, or enhancing tissue and organ functionalities. The artificial scaffolds like framework play a basic role in supporting the structural cells to settle and guide their growth to find the specific tissue with acceptable structure [10]. Therefore, designation of artificial scaffolds has a great importance in tissue engineering. An artificial scaffold that covers the preferred characteristics such as biocompatibility, biodegradability, and high porosity structure could provide as template for bone growth [11]. In the same case, fibrous artificial scaffolds, biodegradable and biocompatible polymers, which are frequently used as artificial scaffold materials, are naturally soft in order to mimic the rigidity of natural tissues [12, 13].

Using fillers as a reinforcing agent is not a new idea in the world. Straws were used to reinforce mud bricks since 4000 BC [14]. Now, fibers made from so many kinds of materials in mesoscale such as glass, boron, silicon carbide, alumina, and especially carbon has been used as fillers in composites. Polymer nanocomposites are combination of a polymer matrix and inclusions that have at least one dimension (i.e., length, width, or thickness) in the nanometer size range (**Figure 1**).

In order to achieve ultimate effective properties, the fabrication of nanoparticle reinforced polymers must be optimized [15, 16]. Nowadays, there are several issues that are not well understood in this area and need more theoretical and experimental researches. However, individual research groups have made significant processing advances for particular nanoparticle-polymer systems, universal guidelines regarding the fabrication of nanocomposites do not exist [17]. This is in part due to the complexity of the polymer chemistry, the lack of detailed models describing the processing conditions, and the large list of parameters (specific to the types of polymer and nanotube under consideration) that can influence the polymer/nanoparticle interaction and impact the effective reinforcement properties.

There are four main system requirements for effective reinforcement. These are a large aspect ratio (1), good dispersion (2), alignment (3), and interfacial stress transfer (4).

Reinforcement of biodegradable and biocompatible polymers is a possible approach to overcome some natural limitations of mentioned polymers such as in adequate mechanical properties, insufficient stiffness, high brittleness, and low toughness [18–22]. Also, some researches were focused on evaluation of the properties of biopolymer blends and copolymers [23–27]. Some kinds of polyesters are widely studied as matrix polymer in biocomposites that reinforced with many kinds of reinforcing fillers for improving their applications. Biopolymers are used to produce harmless fluorescent microparticles for in-vivo material penetration researches.

Biodegradable and biocompatible nanomaterials, because of their properties such as controlled release, low toxicity, and enhanced encapsulation effect, are used frequently as drug delivery systems. Nanotechnology highlighted the impact of nanoencapsulation of various disease-related drugs on biodegradable nanoparticles such as poly(L-lactide-*co*-glycolide) (PLGA), poly lactic acid (PLA), chitosan, gelatin, poly caprolactone, and poly-alkyl-cyanoacrylates [3].

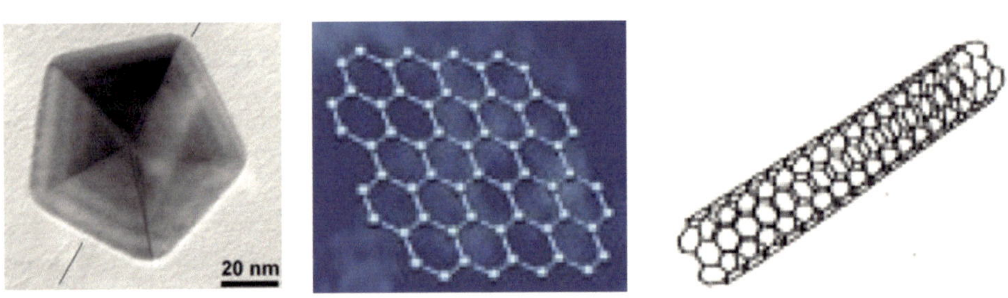

Figure 1.
The scheme of nanofillers for polymer nanocomposites.

2. Polymer nanocomposites fabrication

The methods for fabrication of nanocomposite have considered on improvement of nanomaterials dispersion because significantly higher distribution in the biopolymer matrices to improve the properties of polymeric nanocomposite. Like nanoparticle suspensions in solvents, pristine nanoparticles have not valuable dispersion in polymers illustrating the extreme difficulty to overcome the inherent thermodynamic driving of nanoparticles to agglomerate. The dispersion of nanoparticles in polymer should be evaluated over a various size scales of nanoparticles. The solution blending, melt blending, and in situ polymerization are widely applied to produce nanomaterial/polymer nanocomposites.

2.1 Solution blending of nanocomposites

Solution blending is a common technique for fabrication of polymeric nanocomposites because it is both amenable for various sizes and effectiveness. The solution blending includes three steps: dispersion of nanoparticles in a solvent, mixing with the polymer solution at effective temperature, and finally recover the composite after precipitation or casting the film. Solution-based casting methods provide an advantage through low viscosities, which facilitate mixing and dispersion of the nanoparticles. Many studies have used these methods for processing both thermoset and thermoplastic polymers.

As mentioned earlier, it is difficult to disperse nanoparticles in solvents by simple stirring. The instruments such as ultrasonicator are suitable for making metastable suspensions of reinforcing filler/polymer mixtures in solvents. It is necessary to consider that ultrasonication for a long time affects the nanoparticles. When using solution blending, nanoparticles tend to agglomerate during slow solvent evaporation, leading to inhomogeneous distribution of the nanoparticles in the polymer matrix.

2.2 Melt blending

The melt blending need heat and high shear pressure to disperse the nanoparticles in polymer matrix and it is well-matched with present industries. In comparison with solution blending, the melt mixing has less effective at dispersion of nanoparticles in polymer matrix and has limitation for low concentration of nanoparticle because of high viscosities of the composites at higher nanoparticles loadings.

Melt mixing of nanoparticles into thermoplastic polymer matrix using conventional processing techniques, such as extrusion, injection molding, and blow molding are particularly desirable, due to the speed, simplicity, and availability of the process in plastic industries. These methods are also benefit due to free of solvent and related contaminant. The nanoparticles has a unique advantage in thermoplastic polymer compounding and molding, because less fiber cutting or breaking occurs, and a high aspect ratio is maintained for one dimensional fillers in contrast to larger, microscale fillers. Application of shear mixing with long processing time may improve the dispersion of fillers, and when coupled with elongating extrusion, should yield adequate aligned nanofillers. Increasing in viscosity is higher for nanofibers than that of large diameter fibers such as carbon black, so shear mixing is necessary to overcome the high viscose polymer/nanofibers composites. Additionally, another advantage is the vision of recycling thermoplastic nanocomposites to decrease the financial expenses and to become safe for environment. Nevertheless, much needs to be learned about the ability of nanofibers to withstand high shear and elongation flow processing and about optimization of processing parameters to provide good nanofibers dispersion.

Controlling the alignment of nanofibers in polymer matrix is possible using melt mixing methods. For example, spinning of extruded melt samples is used for alignment of fillers in nanofibers/polypropylene nanocomposites with high dispersion of nanofibers. Up to now, various methods of nanofiber alignment techniques have been developed such as using further increment in residence time in the die channel or die design to control the orientation of nanofibers. Injection molding was also found to induce significant alignment in nanofibers/ polypropylene composites, as demonstrated by measurement of thermal expansion and electrical conductivity.

2.3 In situ polymerization

The nanoparticles are dispersed in monomer and then the polymerization process is starts. As solution blends, functionalized nanoparticles can improve the initial dispersion of the nanoparticles in the liquid (monomer and solvent) and consequently in the nanocomposites. Furthermore, in situ polymerization methods enable covalent bonding between nanoparticles and the polymer matrix using various condensation reactions. Noteworthy extensions of in situ polymerization include infiltration methods in which the reactive agents are introduced into a nanoparticle structure and subsequently polymerized.

3. Biopolymers

A variety of natural, synthetic, and biosynthetic polymers such as poly(L-lactide) (PLLA), polyhydroxyalkanoate (PHA), poly(ε-caprolactone) (PCL), poly glycolic acid (PGA), poly ethylene glycol (PEG), polyesteramide (PEA), and aliphatic copolyesters (PBSA). The biodegradability is capable of undergoing decomposition into carbon dioxide, methane, water, inorganic compounds, or biomass in which the predominant mechanism is the enzymatic action of microorganisms. The biodegradability of polymer depends on the chemical structure of the materials and on the constitution of the final product, and also depends on the raw materials used for its production. The polymer based on the C-C backbone tends to be nonbiodegradable, whereas heteroatom containing polymer backbones confer biodegradability. It is possible to engineer biodegradability of polymers using the judicious addition of chemical linkages such as anhydride, ester, or amide bonds, among others (**Figure 2**). The properties of some commercial biodegradable polymers are summarized in **Table 1**.

The most biomedical application of polymers are surgical dressings, sutures, adhesives, polymeric screws and nails, fiber/polymer composite bone plates, tendons/ligaments, reinforcing meshes, heart valves, joint reconstruction and bone cement, tubular devices, soft-tissue replacement materials for cosmetic reconstruction, drug delivery implants, artificial kidney/blood dialysis, artificial lung/blood oxygenator, and artificial heart.

3.1 Poly(L-lactide)

Among the aliphatic polyesters, poly(L-lactide) (PLLA) is considered to be the most promising biodegradable material, not only because it has excellent biodegradability, compatibility, and high strength but also due to the fact that it can be obtained totally from renewable resources. PLLA is a bio-based, biodegradable polymer which can be produced from renewable sources such as corn and has found numerous applications in the medical and pharmaceutical fields.

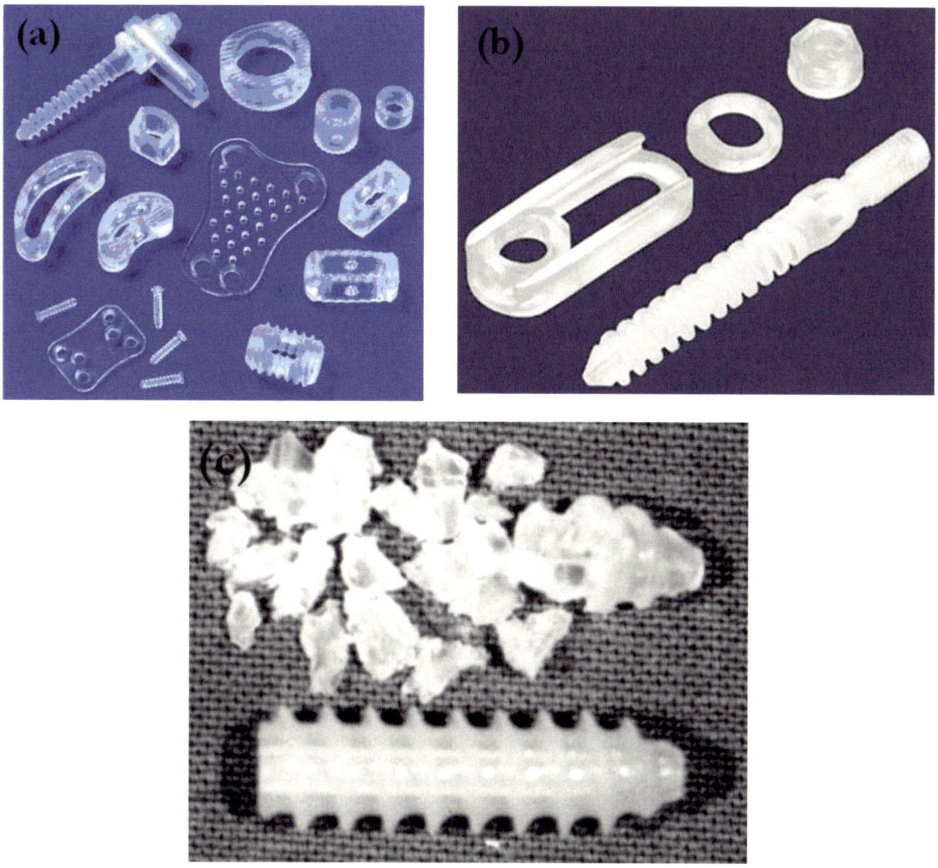

Figure 2.
Bioabsorbable implants that have potential applications throughout the spine (a), an example of a bioabsorbable plate and pedicle screw, washer (b), and the in-vitro degraded bioabsorbable screw.

	PLLA*	PHA*	PCL*	PEA*	PBSA*	PBAT*
Density (g/cm^3)	1.25	1.25	1.11	1.07	1.23	1.21
Melting point (°C) (DSC)	152	153	65	112	114	110–115
Glass transition (°C) (DSC)	58	5	−61	−29	−45	−30
Crystallinity (%)	0–1	51	67	33	41	20–35
Elastic modulus (MPa)	2050	900	190	262	249	52
Elongation at break (%)	9	15	>500	420	>500	>500
Tensile stress at break or max (MPa)	—	—	14	17	19	9
Biodegradation mineralization (%)	100	100	100	100	90	100
Water permeability WVTR at 25°C (g/m^2/day)	172	21	177	680	330	550

**Abbreviations: Poly(L-lactide) (PLLA), polyhydroxyalkanoate (PHA), poly(ε-caprolactone) (PCL), polyesteramide (PEA), aliphatic copolyesters (PBSA), aromatic copolyesters (PBAT).*

Table 1.
Properties of some commercial biopolyesters.

The PLLA has important characteristics over other biopolymers such as:

- Using renewable resources for production,
- Considered as energy saver,
- Recyclable to lactic acid,
- Using carbon dioxide for manufacturing,
- Improving the farm economics by composting,
- Decline of landfill volumes, and
- Possible modification of physical and mechanical properties using copolymerization and blending.

The commercialization of PLLA has been affected from three factors:

- High cost in comparison to other polymers due to its immature technology,
- Moisture absorption of in environment, and
- Modified processing conditions are needed.

Copolymerization of LA with other monomers like glycolide or CL can significantly enhance the properties and broaden the use of PLLA. The PLLA is produced form polylactic acid. The asymmetric polylactic acid has two stereo isomeric forms, L- and D-isomers. The L-isomer exists in normal human carbohydrate metabolism, and the D-isomer is detectable in urine and in acidic milk. If a polymer formed by one type of monomer, it is called homopolymer PLLA. A copolymer consists of two types of monomers named g. poly(D, L)-lactic acid (PDLLA) (**Figure 3**).

Some large scale manufacturers are beginning to favor PLLA because it is renewable, conserves energy, and degrades easily. The ring-opening polymerization of L-lactide oligomers (LAs) yields the PLLA semi-crystalline polymer with a melting point of 180–190°C and a glass transition temperature of 55–60°C (**Figure 4**).

Up to now, the PLLA has limited biomedical applications as implanting devices because of its biodegradation effect. If incorporating different nanoparticles into the PLLA matrix could enhance the properties of this material significantly, this process would increase its applicability further. In addition, the PLLA showed shape memory effect and the original shape could be recovered up to glass transition temperature. However, the recovery strain of PLLA was relatively low and the

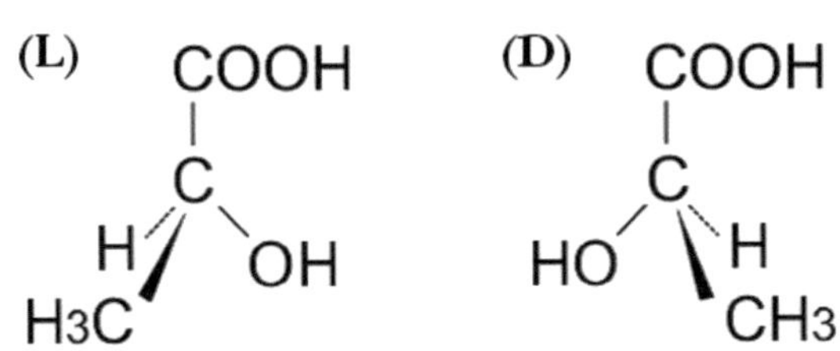

Figure 3.
Stereoisomeric forms of lactic acid: lactic acid occurs in two, L(+) and D(−). Note the difference in location of the hydroxyl group in the chiral carbon.

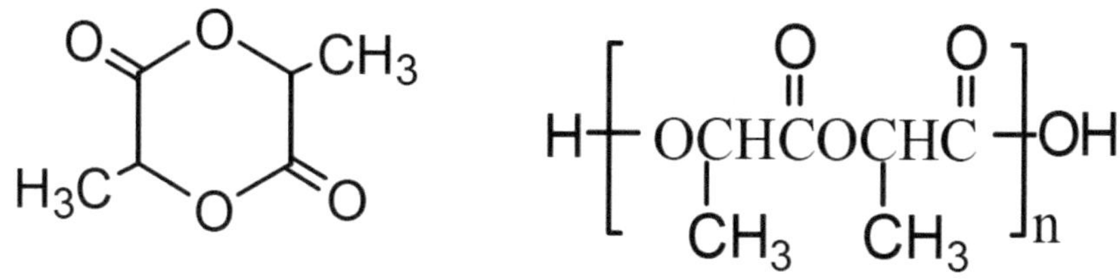

Figure 4.
Molecular scheme of L-lactide as oligomer and poly(L-lactide) as homopolymer.

recovery temperature was high for using in the human body. In order to improve the shape memory and decrease the recovery temperature, copolymers with PCL has been considered.

The PLLA with high performance biodegradable and biocompatible homopolymer is under various studies due to significant properties. PLLA crystallization happens very slowly, even if nanoparticles are incorporated and treat heterogeneous nucleation point [23, 25, 26].

In some researches, the thermo mechanical properties of PLLA nanocomposites reinforced with functionalized multi-walled carbon nanotubes (MWCNT-g-PLLAs) were determined. For functionalization, PLLA chains were grafted form the surface of MWCNTs. Then, the func.MWCNTs/PLLA composite is prepared by solution casting. The results show that the MWCNT-g-PLLAs were dispersed in PLLA matrix adequately. With increasing the weight percentage of MWCNT-g-PLLAs, up to 2 wt% led to gradual enhancement of the mechanical properties of nanocomposite. The thermal analysis also revealed the func.MWCNTs increase the melting point and the glass transition temperature of nanocomposite. Also, the DMA analysis results show that incrementing the concentrations of func.MWCNTs is also accompany with increasing Young modulus and the transition temperature of PLLA. The chain stiffness in amorphous phase of PLLA can also increase due to the van der Walls force and the homogenous dispersion of func.MWCNTs. In addition, the crystallinity of PLLA could be increased due to func.MWCNTs as heterogeneous nucleation points [19, 28].

3.2 Poly(L-lactide-co-ε-caprolactone)

Poly(ε-caprolactone) (PCL) is another important aliphatic polyester that is considered as a potential material in both biomedical and environmental fields. PCL is a biodegradable and nontoxic polyester. The ring-opening polymerization of ε-caprolactone oligomers (CLs) yields the PCL semi-crystalline polymer with a melting point of 59–64°C and a glass transition temperature of −60°C. The glass transition temperature can be increased by copolymerization with L-lactide, which also enhances the biodegradation of the polymer. PCL has good permeability to many therapeutic drugs and has been studied for long-term contraceptive delivery (**Figure 5**).

The polymer has been regarded as tissue compatible and used as a biodegradable suture. PCL exhibits a low glass transition temperature and melting point, high crystallinity and permeability, and good flexibility with a high elongation at break and low modulus. However, modification is highly necessary when it is applied to different requirements. Because the homopolymer has a degradation time on the order of 2 years, copolymers have been synthesized to accelerate the rate of bioabsorption.

For example, copolymers of CL with LA have yielded materials with more rapid degradation rates. Also, combining nanoparticles with PCL is an effective and operable approach to improving the properties of PLLA significantly. The copolymers of PLLA with other biopolymers such as PCL may increase its applications because with this procedure, it becomes possible to fabricate a various kinds of bioabsorbable polymers and composites with soft and elastic properties. Because the PCL has a low melting point, if PCL is introduced into segmented polyurethane as a soft segment, the shape memory effect would be expected. Hydrolysis of PCL yields 6-hydroxycaproic acid which enters the citric acid cycle and is metabolized.

PLLA is a biocompatible and biodegradable homopolymer with good mechanical properties and its copolymers with PCL may expand its applications. The CL appears to be a suitable comonomer for the preparation of copolymers with PLLA and PGA with mechanical properties ranging from rigid to elastomeric. The copolymer of PLLA and PCL possessed properties partly like that of PLLA and partly like that of PCL (**Figure 6**).

The poly(L-lactide-co-ε-caprolactone) PLACL has a lower tensile strength than higher elongation and substantially more rapid degradation time than PLLA. But PLACL has not enough sufficient characters for hard tissue engineering. The synthesis of LA/CL copolymers and other lactone polymers have been widely studied in recent years. Most studies have focused on random, diblock, and triblock copolymers. Both PLLA and PCL have shape memory properties. Hence the PLACL must have shape memory effect. It is found that the mechanical properties of PLACL are significantly affected by the polymer compositions. With the increment of CL content, the maximum stress decreases linearly and the strain at break increases gradually as can be seen in **Figure 6**. By adjusting the compositions of monomers, the copolymers exhibit excellent shape memory effects.

There are many research on reinforcing the PLACL using nanomaterials. As an example, PLACL reinforced with well-dispersed multi-walled carbon nanotubes (MWCNTs) were prepared using functionalized MWCNT by in situ polymerization. The surface functionalization of MWCNTs can effectively improve the

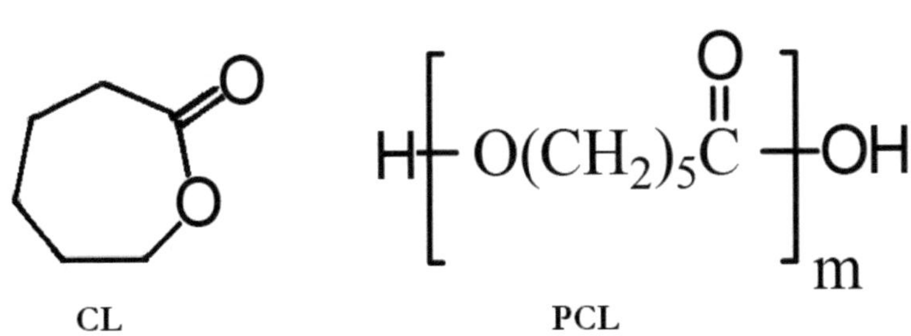

Figure 5.
Molecular scheme of ε-caprolactone as oligomer and poly(ε-caprolactone) as homopolymer.

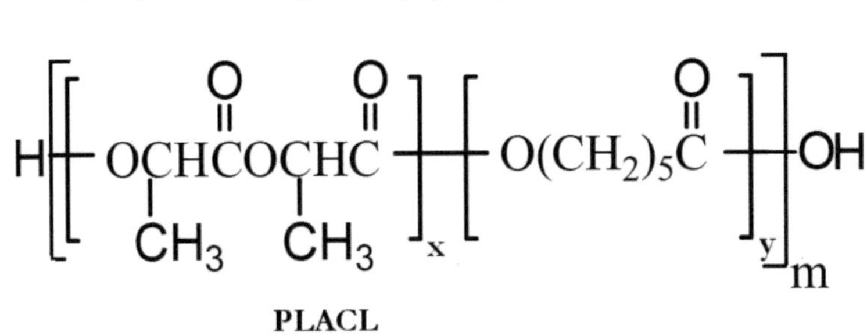

Figure 6.
Molecular scheme poly(L-lactide-co-ε-caprolactone) as copolymer.

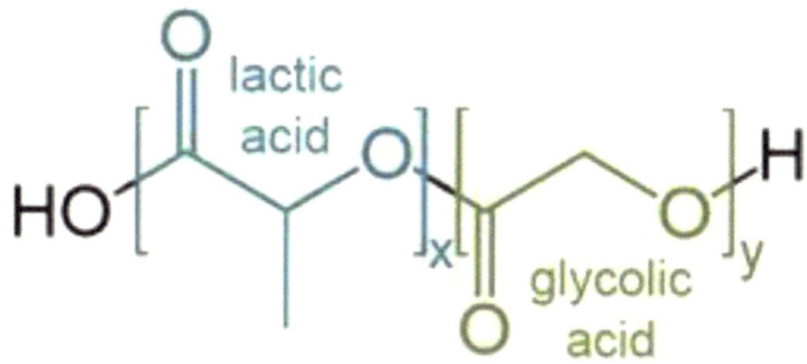

Figure 7.
Molecular scheme poly(D,L-lactide-co-glycolide) as copolymer.

dispersion and adhesion of MWCNTs in PLACL and hence, it will have a significant effect on the physical, thermomechanical, and degradation properties of MWCNT/PLACL nanocomposites [29].

3.3 Poly(D,L-lactide-co-glycolide) (PLGA)

The next nanocarrier that has been considered for sustained and targeted delivery of different agents is poly[L-lactide-co-glycolide] (PLGA)-based nanoparticles. Although PLGA have been applied many years ago, but the task of nanoparticles in mechanism of intercellular uptake, their trafficking, and sorting into different intercellular compartments, as well as their procedure of action for therapeutic efficacy of nanoparticles encapsulated agent at cellular level is recently considered [30] (**Figure 7**).

In addition, we know that the PLGA nanoparticles have deeper in vitro and in vivo effects in comparison with industrial nanoparticles in similar range such as ferrous oxide and zinc oxide. The effect of PLGA nanoparticles on cell viability was characterized by in vitro cytotoxicity analysis via a WST assay. The PLGA, silica, and ferrous oxide have a cell viability up to 75%, but for zinc oxide, particles cell viability significantly reduced [31]. The researchers found that nanoparticle mean size correlates linearly with polymer concentration is between 70 and 250 nm [32].

The PLGA/MWCNT composite was considered as a scaffold material to treat artificial bloods. PLGA/MWCNT nanocomposite is prepared using electrostatic technique, in which layers of MWCNTs are deposited on the PLGA. For in vivo and in vitro analysis, the fibrinogen is immobilized on PLGA/MWCNT composite and incubated in non-stimulated platelet-rich plasma (PRP) for platelet studies. The interaction of fibrinogen and PRP, are characterized on the prepared PLGA/MWCNT nanocomposite [33].

4. Nanoparticles

Nanomaterials consists of materials that the size of particle is less than 100 nm. All kinds of materials could be treating to be nanomaterials such as metallic, nonmetallic, ceramics, polymeric and so on..

4.1 Metal-based nanoparticles

The widely used metallic nanoparticles in the field of medicine and biotechnology are gold (Au), platinum (Pt), silver (Ag), selenium (Se), copper (Cu), palladium (Pd), and gadolinium (Gd), also, the widely used metal oxide nanoparticles in the field of medicine and biotechnology are iron oxide (Fe_2O_3), titanium oxide (TiO_2), and zinc oxide (ZnO). The metallic nanomaterials can be prepared and modified with appropriate chemical functional groups to bind with drugs, antibodies, and ligands.

4.1.1 Gold-based nanomaterials (AuNPs)

The AuNPs have many characteristics such as biocompatibility, optical properties, and electrical behavior. Now–a-days, AuNPs have been considered in bioimaging and tissue engineering. Electrospinning and metal nanoparticles (Nps) can create a scaffold that will trigger muscle cell elongation, orientation, fusion, and striation. Traumatic injuries can interrupt muscle contraction by damaging the skeletal muscle and/or the peripheral nerves. The healing process results in scar tissue formation that impedes muscle function. Poly(L-lactic acid) (PLLA) and Nps were electrospun to create nanocomposite by Fischer et al. [34]. They found that low amounts of AuNps may be utilized to create a biodegradable, biocompatible, and conductive scaffold for skeletal muscle repair.

4.1.2 Silver-based nanomaterials (AgNPs)

Silver nanoparticles have antimicrobial activity and useful as antimicrobial agent, hence, it is a proven killer of bacteria [35]. Silver is far more efficient antibiotic than any allopathic pharmaceutical materials. Colloidal silver is effective in killing more than 600 bacteria in less than 5 min. AgNPs also find application in ointment and cream used to prevent infection in burns and open wounds anticancer particles with paclitaxel inhibits the growth of hep G2 cell more effectively [36–38].

Biodegradable PLLA ultrafine fibers containing AgNps were prepared via electrospinning by Xu et al. [39]. These fibers showed antibacterial activities (microorganism reduction) of 98.5 and 94.2% against *Staphylococcus aureus* and *Escherichia coli*, respectively, because of the presence of the silver nanoparticles.

4.1.3 Copper-based nanomaterials (CuNPs)

The polymer/CuNPs loading is proposed as a biostatic coating and systematic correlations between material properties and biological effects are established. The experimental result of the nanocomposite capability to release metals in a controlled manner and to slow or inhibit the growth of living organisms are proofed [40].

Using the electrospinning method, Badaraev et al. produced biodegradable scaffolds from PLLA. Using DC magnetron sputtering of the copper target, they modified the surface of the scaffolds. The diameters of fibers range from 0.8 to 2 μm. Testing for antibacterial features indicated that the modified scaffolds are capable to have a bacteriostatic effect [41].

4.1.4 Selenium-based nanomaterials (SeNPs)

The major biomedical applications of SeNPs include, targeted drug delivery [42–44], drug delivery vehicles and artificial enzymes [45, 46], anti-cancer therapy [47–49], anti-bacterial activities [50], biosensors and intracellular analysis [51].

For bone tissue engineering, application of bioactive glass scaffolds because of bone bonding ability is present interests. Of course the bioactive glass scaffolds do not have some functionalities to enable the successful formation of new bone. For bone tissue engineering, application of Se due to significant role in antioxidant protection enhanced immune surveillance and modulation of cell proliferation is a solution for problem. Also, the SeNPs possess antibacterial as well as antiviral activities. Stevanović et al., in their recent research, synthesized uniform, stable, amorphous SeNPs, and additionally immobilized within spherical PLGA particles (PLGA/SeNPs). These particles were used to coat bioactive glass-based scaffolds synthesized by the foam replica method. The prepared composite showed a considerable antibacterial activity against Gram-positive bacteria, *Staphylococcus aureus* and *Staphylococcus epidermidis*, one of the main causative agents of orthopedic infections [52].

4.1.5 Palladium-based nanomaterials (PdNPs)

The major biomedical applications of PdNPs include targeted drug delivery [53, 54], anti-cancer therapy [55, 56], anti-microbial activities [57], biosensors and intracellular analysis-hydrogen sensors [58, 59], biocatalysts [60], and catalysis [61]. Graphene oxide (GO) has treated to create an anchoring OH site on the surface of GO. The subsequent GO-g-PLA was synthesized by the polymerization reaction in the presence of GO-MDI-OH and PLA. Finally, GO-g-PLA-Pd NPs was used for the electrochemical detection of serotonin [62].

4.2 Metal oxides-based nanomaterials

Biodegradation and biocompatibility of metal oxide nanoparticles (MONPs) are investigated medical applications. It is vital that the surface modification of MOPs must be adequate stable to resist against the salts and proteins in vivo and also become water soluble. It is elucidated that super paramagnetic iron oxides nanoparticles (SPIONPs) are significantly biocompatible. The behavior of SPIONPs for drug delivery applications based on their surface structure and conjugated targeting ligands/proteins [63–66].

The most important application of SPIONPs are include targeting of drug by engineered delivering system [67, 68], for cancer therapeutic [69, 70], diagnosis of many kinds of cancers [67], contrasting agents for bioimaging [71], ultra-sensitive in vivo molecular imaging [72], anti-microbial activities [73, 74], biosensing and inter cellular analysis [75], and cancer therapy using photo thermal technique [76, 77]. The distinctive properties of iron oxide MNPs are appropriate for biocatalysis [78, 79].

ZnO NPs are used as anti-microbial, anti-biotic, and anti-fungal (fungicide) agents by incorporating them in coatings, bandages, nanofiber, nanowire, plastics, alloy, and textiles. They possess suitable electrical, dielectric, magnetic, optical, imaging, catalytic, biomedical, and bioscience properties. ZnO is a white powder that insoluble in water. ZnO is applicable in many kinds of ointments that used to treat skin irritations. Also, ZnO has many industrial applications such as in semiconductors, ceramics, and glass compositions [80, 81]. The well-known biomedical applications of ZnO NPs are found as targeted drug delivery destruction of tumor cells [82, 83], biomedical imaging and drug delivery systems [84], tumor characterization [85, 86], anti-cancer therapy [87], contrast agent in medical imaging [88], anti-microbial activities [89], biomarkers [90], and biosensors [91–94].

A suitable food packaging can increase the shelf life of food products in addition to save their initial quality. The biodegradable polymer has various limitations such as fragility due to their low mechanical properties. Due to high aspect ratio of nanoparticles, their properties have significant differences from conventional size particles. ZnO nanostructured materials have presented valuable properties which have led to variety of applications such as food packaging applications.

Combination of ZnO nanoparticles and polyvinyl alcohol results a more effective and environmentally friendly material for food stuff packaging [95].

Titanum oxide (TiO_2) nanoparticles can enhance cell attachment and proliferation on its composite surfaces. The polymer/TiO_2 composite films exhibit enhanced cell adhesion and a tendency to increased Ca-containing mineral deposition. Also, TiO_2 nanoparticles might act as interfacial bonding to tissue by means of the formation of a biologically active hydroxyapatite layer on implant surface. Boccaccini fabricated PDLLA films contain TiO_2 nanoparticles. Thus, if TiO_2 NPs are introduced in PLLA matrix, some disadvantages are anticipated to be improved. However, one of the most problems in master batch production of TiO_2 is the agglomeration in the PLLA matrix. The aggregated TiO_2 NPs in the composite reduce the mechanical properties and hence is necessary more researches to solve the TiO_2 agglomeration [96].

Deterioration of fresh fruits and vegetables during storage treat microorganisms breeding such as *Aspergillus niger* (*A. niger*) and *Bacillus subtilis* (*B. subtilis*), which can be a seriously danger for human health. For antibacterial and preservative properties, a self-assembled film of graphene oxide (GO) and chitosan (CS) with titanium dioxide (TiO_2) nanoparticles are introduced. These non-cytotoxic nanometer-scale films, with the ratio of 1:20:4 for graphene oxide, chitosan, and titanium dioxide nanoparticles, respectively, exhibited valuable antibacterial activity against the biofilm forming strains *A. niger* and *B. subtilis*. Also, the nanocomposites did not show any cytotoxicity against mammalian somatic cells and plant cells. Nanocomposites disrupted microbial film formation while avoiding internalization by animal and plant cells. Due to their selectivity and safety, these nanocomposites demonstrate potential as antimicrobial coatings for food preservation [97].

4.3 Silica nanoparticles (SNPs)

The performance of SNPs as nanofillers in polymer nanocomposite has significant attention, because of increased in demand for new materials with enhancement in thermal, mechanical, physical, and chemical properties of various kinds of composites (**Figure 8**). Synthesis of SNPs using sol-gel treatment has significant improvement in the development of SNPs/polymer nanocomposites [98].

Gardella et al. developed a novel catalytic system, consisting of palladium nanoclusters homogenously dispersed on the surface of nanostructured polymer fibers based on poly(L-lactide) (PLLA) and polyhedral oligomeric silsesquioxanes (POSS). In fact, PLLA nanofibers that contain amino silsesquioxane molecules (POSS-NH_2) have capability to interact with metal precursor prepared by electrospinning. The prepared system proves a relevant catalytic activity toward the hydrogenation of stilbene under heterogeneous conditions [99].

4.4 Carbon-based nanomaterials

4.4.1 Carbon nanotubes

The carbon nanotubes (CNTs) have been investigated for a variety of applications based on their unique electrical, optical, and mechanical properties. The exceptional mechanical properties of CNTs have led to their use as effective reinforcing filler for polymer composites. It was expected that CNTs would display superlative mechanical properties by analogy with graphite. The inside of CNTs can be filled with some elements or compounds, such as C_{60}, to produce hybrid nanomaterials which possess unique intrinsic properties. The properties of the CNTs/

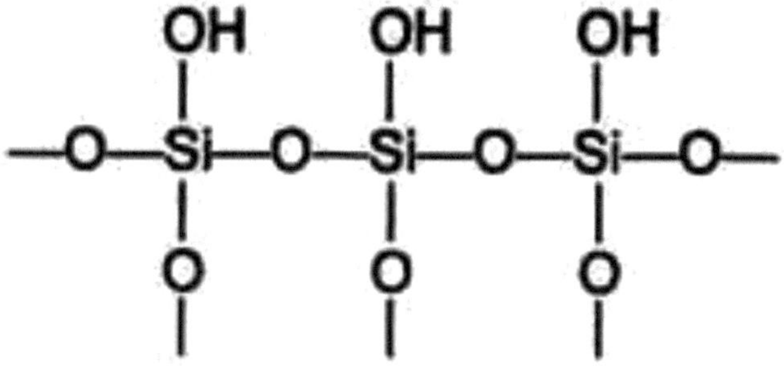

Figure 8.
Chemical scheme of silica.

polymer composites will vary significantly depending on the distribution of the type, diameter, and length of the nanotubes.

In CNTs, only few concentrated acids are capable of breaking the bonds between carbon atoms. Consequently, when CNTs reinforce a composite, the mentioned stability becomes a problem at the interaction between the matrix and CNTs. Uncontrolled agglomeration is another noticeable difficulty that can interfere on CNTs due to its nanometer size. To increase the interaction between matrix and reinforcement is submitting CNTs to a process called functionalization. Functionalization of CNTs is a mix of physical and chemical processes that inserts functional groups on the sidewall of CNTs. The introduction of this procedure can also be helpful to obtain better dispersion of carbon nanotubes into relevant matrices (**Figure 9**).

4.4.2 Functionalization of CNTs

Functionalization is one of the most effective methods in improving the surface properties of CNTs so that the application potentials can be fully realized. The methods of functionalization for CNTs range from chemical modification to physical interaction, and mechanical manipulations (**Figure 10**).

4.4.3 Functionalization in chemical base

In chemical base, there are two methods for functionalization of CNTs:

1. Covalent functionalization.

2. Noncovalent functionalization.

Each method has some advantages and some disadvantages.

The carbon nanotubes (CNTs) are another important and novel category of NPs that has been investigated extensively in medicine and drug delivery systems. The CNTs can interact with various bio-macromolecules such as DNA and proteins by physical adsorption. Additionally, in order to conjugate covalently targeting moieties or therapeutic molecules to CNTs, numerous chemical modifications were developed [117, 118].

In the field of research on medical application of CNTs, Zheng et al. elucidated the interactions between DNA molecules and CNTs [119]. In the case of single-stranded DNA, CNTs could disperse effectively in aqueous media. Up to date, the improvement of mechanical properties of CNTs might be counted primarily for

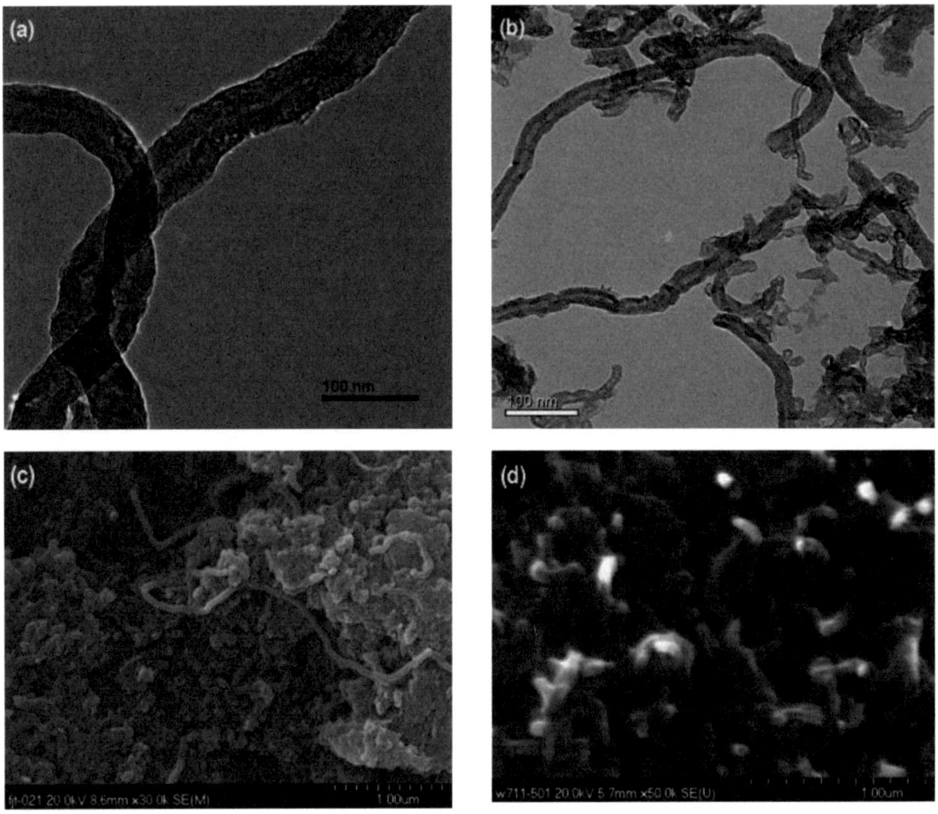

Figure 9.
*TEM micrograph of pristine MWCNTs (a), MWCNT-COOHs (b), MWCNT-OHs (c, d), and MWCNT-OH-*graft*-PLACL(e, f) [100].*

their using as composite reinforcements for tissue engineering and preparation of artificial scaffolds [120]. More recently, researchers have considered their attention to utilizing the multi-functional nature of CNTs in engineering tissue scaffolds. Most particularly, the CNTs were incorporated to fabricate electrically conductive artificial scaffolds.

4.4.4 Graphene nanoparticles

Due to the similarity between graphene and CNTs, several medical applications such as drug delivery systems, scaffold reinforcements, and injectable cellular labeling agents have been committed using graphene and graphene oxide (GO) [121]. For reinforcement of biodegradable polymers by graphene, in one case, the PLLA/GO nanocomposites were prepared by solution mixing. The results show that the crystallization of PLLA enhanced and the spherulite morphology change were insignificant when the content of GO exceeded 0.5 wt%, because the extreme GO increased the number of nucleation sites while restricting the PLA crystal growth. Thus, the arrangement of nanopores did not mimic the spherulites because of imperfect crystal morphology [122].

4.5 Nano hydroxyapatite (HANs)

Owing to its biocompatibility and osteoconductive properties, nano hydroxyapatite (nHA) is widely used bioceramic for bone graft substitute. nHA with

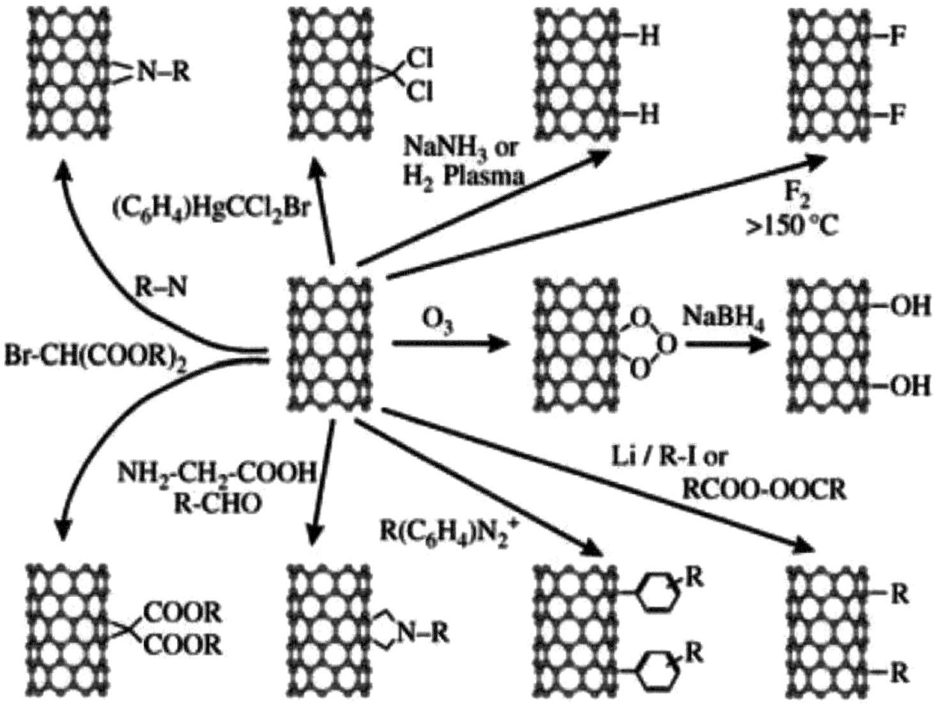

Figure 10.
Scheme of possible addition reactions for the functionalization of CNTs [101–116].

biodegradable and biocompatible polymer-based composite scaffolds have been explored for bone grafting. Hence, the nHA/biopolymer nanocomposites proved to be promising for bone tissue engineering [123]. Composite fibers composed of PLA-g-HANs and PLA matrix was prepared by electro-spinning for tissue engineering [124].

4.6 Magnetic nanoparticles (MNs)

The MNs are a class of nanomaterials which can be performed using adequate magnetic field. The MNs can be conjugated with any protein, drug and gene, and by that MNs serve as contrast agent for magnetic resonance imaging (MRI) by changing the MRI signal. Additionally, MNs serve as a therapeutic tool by improving drug delivery to the target organ. Drug controlled releasing using nanostructured functional materials are attracting increasing attention in some diseases such as cancer therapy and other ailments. The potential of MNs stems from the intrinsic properties of their magnetic cores combined with their drug loading capability and the biochemical properties [125]. Therapeutic compounds are attached to MNs and magnetic fields generated outside the body are focused on specific targets [126].

In the field of biopolymer nanocomposites, iron oxide MNs with sizes less than 10 nm have been successfully deposited on multi-walled CNTs (MWCNTs) by in situ high temperature decomposition of iron(III) acetylacetonate and MWCNTs in polyol solution [127]. The PLLA has been covalently grafted onto the surface of mMWCNTs. The mMWCNTs/PLLA nanocomposite possess significant mechanical, electronic, super paramagnetic, and biocompatible properties, which means that the mMWCNTs/PLLA will have great potential applications in the fields of nanobiomaterials and nanotechnology, and the addition of mMWCNTs/PLLA can treat novel properties to PLLA and other biodegradable polymers [128] (**Figure 11**).

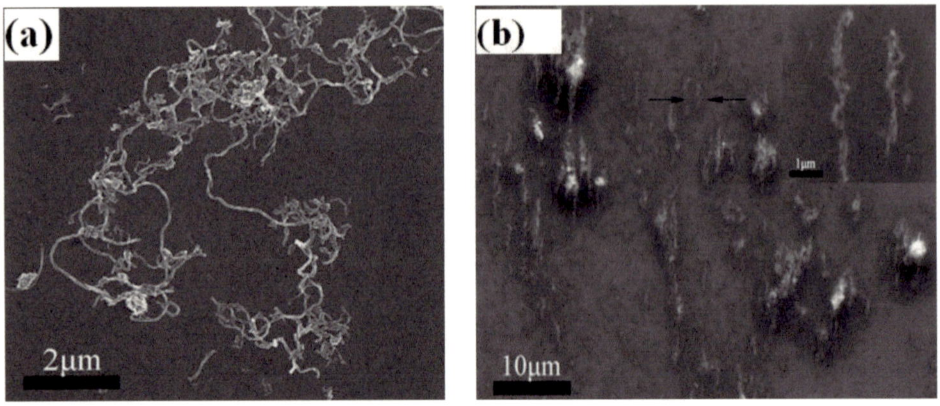

Figure 11.
SEM images of m-MWCNTs-g-PLLA in the absence (a) and presence (b) of an external magnetic field [128].

Author details

Ali Nabipour Chakoli
Nuclear Science and Technology Research Institute, Tehran, Iran

*Address all correspondence to: anabipour@aeoi.org.ir

References

[1] Valavanidis A, Vlachogianni T. Engineered nanomaterials for pharmaceutical and biomedical products new trends, benefits and opportunities. Pharmaceutical Bioprocessing. 2016;**4**(1):13-24

[2] Palza H. Antimicrobial polymers with metal nanoparticles. International Journal of Molecular Sciences. 2015;**16**(1):2099-2116

[3] Kumari A, Yadav SK, Yadav SC. Biodegradable polymeric nanoparticles based drug delivery systems. Colloids and Surfaces B: Biointerfaces. 2010;**75**(1):1-18

[4] Armentano I, Dottori M, Fortunati E, Mattioli S, Kenny J. Biodegradable polymer matrix nanocomposites for tissue engineering: A review. Polymer Degradation and Stability. 2010;**95**(11):2126-2146

[5] Okamoto M, John B. Synthetic biopolymer nanocomposites for tissue engineering scaffolds. Progress in Polymer Science. 2013;**38**(10-11):1487-1503

[6] Nitta S, Numata K. Biopolymer-based nanoparticles for drug/gene delivery and tissue engineering. International Journal of Molecular Sciences. 2013;**14**(1):1629-1654

[7] Van Vlierberghe S, Dubruel P, Schacht E. Biopolymer-based hydrogels as scaffolds for tissue engineering applications: A review. Biomacromolecules. 2011;**12**(5):1387-1408

[8] Li X, Cui R, Sun L, Aifantis KE, Fan Y, Feng Q, et al. 3D-printed biopolymers for tissue engineering application. International Journal of Polymer Science. 2014;**2014**:3. Article ID: 829145. DOI: 10.1155/2014/829145

[9] Pina S, Oliveira JM, Reis RL. Natural-based nanocomposites for bone tissue engineering and regenerative medicine: A review. Advanced Materials. 2015;**27**(7):1143-1169

[10] Wang X, Jiang M, Zhou Z, Gou J, Hui D. 3D printing of polymer matrix composites: A review and prospective. Composites Part B: Engineering. 2017;**110**:442-458

[11] Mogoşanu GD, Grumezescu AM. Natural and synthetic polymers for wounds and burns dressing. International Journal of Pharmaceutics. 2014;**463**(2):127-136

[12] Guo B, Ma PX. Synthetic biodegradable functional polymers for tissue engineering: A brief review. SCIENCE CHINA Chemistry. 2014;**57**(4):490-500

[13] Liu X, Holzwarth JM, Ma PX. Functionalized synthetic biodegradable polymer scaffolds for tissue engineering. Macromolecular Bioscience. 2012;**12**(7):911-919

[14] Coleman JN, Khan U, Blau WJ, Gun'ko YK. Small but strong: A review of the mechanical properties of carbon nanotube–polymer composites. Carbon. 2006;**44**(9):1624-1652

[15] Sahoo NG, Rana S, Cho JW, Li L, Chan SH. Polymer nanocomposites based on functionalized carbon nanotubes. Progress in Polymer Science. 2010;**35**(7):837-867

[16] Murphy EB, Wudl F. The world of smart healable materials. Progress in Polymer Science. 2010;**35**(1-2):223-251

[17] Kumar SK, Jouault N, Benicewicz B, Neely T. Nanocomposites with polymer grafted nanoparticles. Macromolecules. 2013;**46**(9):3199-3214

[18] Amirian M, Chakoli AN, Sui JH, Cai W. Enhanced mechanical and photoluminescence effect of poly (l-lactide) reinforced with functionalized multiwalled carbon nanotubes. Polymer Bulletin. 2012;**68**(6):1747-1763

[19] Amirian M, Chakoli AN, Sui JH, Cai W. Thermo-mechanical properties of MWCNT-g-poly (l-lactide)/poly (l-lactide) nanocomposites. Polymer Bulletin. 2013;**70**(10):2741-2754

[20] Amirian M, Nabipour Chakoli A, Cai W, Sui J. Effect of functionalized multiwalled carbon nanotubes on thermal stability of poly (L-LACTIDE) biodegradable polymer. Scientia Iranica. 2013;**20**(3):1023-1027

[21] Amirian M, Nabipour Chakoli A, Zamani Zeinali H, Afarideh H. Enhanced photoluminescence effect of poly (L-lactide) biodegradable polymer with functionalized carbon nanotubes. Advanced Materials Research. 2014;**829**:304-308

[22] Chakoli AN, He J, Cheng W, Huang Y. Enhanced oxidized regenerated cellulose with functionalized multiwalled carbon nanotubes for hemostasis applications. RSC Advances. 2014;**4**(94):52372-52378

[23] Hong Z, Zhang P, He C, Qiu X, Liu A, Chen L, et al. Nano-composite of poly(L-lactide) and surface grafted hydroxyapatite: Mechanical properties and biocompatibility. Biomaterials. 2005;**26**(32):6296-6304

[24] Xu J-Z, Chen T, Yang C-L, Li Z-M, Mao Y-M, Zeng B-Q, et al. Isothermal crystallization of poly (l-lactide) induced by graphene nanosheets and carbon nanotubes: A comparative study. Macromolecules. 2010;**43**(11):5000-5008

[25] Zhang D, Kandadai MA, Cech J, Roth S, Curran SA. Poly (L-lactide) (PLLA)/multiwalled carbon nanotube (MWCNT) composite: Characterization and biocompatibility evaluation. The Journal of Physical Chemistry B. 2006;**110**(26):12910-12915

[26] Zhang P, Hong Z, Yu T, Chen X, Jing X. In vivo mineralization and osteogenesis of nanocomposite scaffold of poly (lactide-co-glycolide) and hydroxyapatite surface-grafted with poly (L-lactide). Biomaterials. 2009;**30**(1):58-70

[27] Zhou WY, Lee SH, Wang M, Cheung WL, Ip WY. Selective laser sintering of porous tissue engineering scaffolds from poly (L-lactide)/carbonated hydroxyapatite nanocomposite microspheres. Journal of Materials Science: Materials in Medicine. 2008;**19**(7):2535-2540

[28] Chakoli AN, He J, Chayjan MA, Huang Y, Zhang B. Irradiation of poly (L-lactide) biopolymer reinforced with functionalized MWCNTs. RSC Advances. 2015;**5**(68):55544-55549

[29] Amirian M, Sui J, Chakoli AN, Cai W. Properties and degradation behavior of surface functionalized MWCNT/poly (L-lactide-co-ε-caprolactone) biodegradable nanocomposites. Journal of Applied Polymer Science. 2011;**122**(5):3133-3144

[30] Panyam J, Labhasetwar V. Biodegradable nanoparticles for drug and gene delivery to cells and tissue. Advanced Drug Delivery Reviews. 2003;**55**(3):329-347

[31] Semete B, Booysen L, Lemmer Y, Kalombo L, Katata L, Verschoor J, et al. In vivo evaluation of the biodistribution and safety of PLGA nanoparticles as drug delivery systems. Nanomedicine: Nanotechnology, Biology and Medicine. 2010;**6**(5):662-671

[32] Cheng J, Teply BA, Sherifi I, Sung J, Luther G, Gu FX, et al. Formulation

of functionalized PLGA–PEG nanoparticles for in vivo targeted drug delivery. Biomaterials. 2007;**28**(5):869-876

[33] Koh LB, Rodriguez I, Zhou J. Platelet adhesion studies on nanostructured poly(lactic-co-glycolic-acid)–carbon nanotube composite. Journal of Biomedical Materials Research Part A. 2008;**86**(2):394-401

[34] McKeon-Fischer K, Freeman J. Characterization of electrospun poly(L-lactide) and gold nanoparticle composite scaffolds for skeletal muscle tissue engineering. Journal of Tissue Engineering and Regenerative Medicine. 2011;**5**(7):560-568

[35] Gnanadesigan M, Anand M, Ravikumar S, Maruthupandy M, Ali MS, Vijayakumar V, et al. Antibacterial potential of biosynthesised silver nanoparticles using *Avicennia marina* mangrove plant. Applied Nanoscience. 2012;**2**(2):143-147

[36] Zuco V, Supino R, Righetti SC, Cleris L, Marchesi E, Gambacorti-Passerini C, et al. Selective cytotoxicity of betulinic acid on tumor cell lines, but not on normal cells. Cancer Letters. 2002;**175**(1):17-25

[37] Jeyaraj M, Sathishkumar G, Sivanandhan G, MubarakAli D, Rajesh M, Arun R, et al. Biogenic silver nanoparticles for cancer treatment: An experimental report. Colloids and Surfaces B: Biointerfaces. 2013;**106**:86-92

[38] Ong C, Lim J, Ng C, Li J, Yung L, Bay B. Silver nanoparticles in cancer: Therapeutic efficacy and toxicity. Current Medicinal Chemistry. 2013;**20**(6):772-781

[39] Xu X, Yang Q, Wang Y, Yu H, Chen X, Jing X. Biodegradable electrospun poly (L-lactide) fibers containing antibacterial silver nanoparticles. European Polymer Journal. 2006;**42**(9):2081-2087

[40] Cioffi N, Torsi L, Ditaranto N, Tantillo G, Ghibelli L, Sabbatini L, et al. Copper nanoparticle/polymer composites with antifungal and bacteriostatic properties. Chemistry of Materials. 2005;**17**(21):5255-5262

[41] Badaraev A, Nemoykina A, Bolbasov E, Tverdokhlebov S. PLLA scaffold modification using magnetron sputtering of the copper target to provide antibacterial properties. Resource-Efficient Technologies. 2017;**3**(2):204-211

[42] Yu B, Zhang Y, Zheng W, Fan C, Chen T. Positive surface charge enhances selective cellular uptake and anticancer efficacy of selenium nanoparticles. Inorganic Chemistry. 2012;**51**(16):8956-8963

[43] Liu N, Wang Z, Ma Z. Platinum porous nanoparticles for the detection of cancer biomarkers: What are the advantages over existing techniques? Bioanalysis. 2014;**6**(7):903-905

[44] Wang Z, Liu N, Ma Z. Platinum porous nanoparticles hybrid with metal ions as probes for simultaneous detection of multiplex cancer biomarkers. Biosensors and Bioelectronics. 2014;**53**:324-329

[45] Liu T, Zeng L, Jiang W, Fu Y, Zheng W, Chen T. Rational design of cancer-targeted selenium nanoparticles to antagonize multidrug resistance in cancer cells. Nanomedicine: Nanotechnology, Biology and Medicine. 2015;**11**(4):947-958

[46] Sun D, Liu Y, Yu Q, Qin X, Yang L, Zhou Y, et al. Inhibition of tumor growth and vasculature and fluorescence imaging using functionalized ruthenium-thiol protected selenium nanoparticles. Biomaterials. 2014;**35**(5):1572-1583

[47] Xia Y, You P, Xu F, Liu J, Xing F. Novel functionalized selenium nanoparticles for enhanced anti-hepatocarcinoma activity in vitro. Nanoscale Research Letters. 2015;**10**(1):349

[48] Zheng W, Cao C, Liu Y, Yu Q, Zheng C, Sun D, et al. Multifunctional polyamidoamine-modified selenium nanoparticles dual-delivering siRNA and cisplatin to A549/DDP cells for reversal multidrug resistance. Acta Biomaterialia. 2015;**11**:368-380

[49] Xu H, Cao W, Zhang X. Selenium-containing polymers: Promising biomaterials for controlled release and enzyme mimics. Accounts of Chemical Research. 2013;**46**(7):1647-1658

[50] Wang X, Sun K, Tan Y, Wu S, Zhang J. Efficacy and safety of selenium nanoparticles administered intraperitoneally for the prevention of growth of cancer cells in the peritoneal cavity. Free Radical Biology and Medicine. 2014;**72**:1-10

[51] Yang F, Tang Q, Zhong X, Bai Y, Chen T, Zhang Y, et al. Surface decoration by Spirulina polysaccharide enhances the cellular uptake and anticancer efficacy of selenium nanoparticles. International Journal of Nanomedicine. 2012;**7**:835

[52] Stevanović M, Filipović N, Djurdjević J, Lukić M, Milenković M, Boccaccini A. 45S5Bioglass®-based scaffolds coated with selenium nanoparticles or with poly (lactide-co-glycolide)/selenium particles: Processing, evaluation and antibacterial activity. Colloids and Surfaces, B: Biointerfaces. 2015;**132**:208-215

[53] Prasad KS, Vaghasiya JV, Soni SS, Patel J, Patel R, Kumari M, et al. Microbial selenium nanoparticles (SeNPs) and their application as a sensitive hydrogen peroxide biosensor. Applied Biochemistry and Biotechnology. 2015;**177**(6):1386-1393

[54] Tran PA, Webster TJ. Selenium nanoparticles inhibit *Staphylococcus aureus* growth. International Journal of Nanomedicine. 2011;**6**:1553

[55] Abu-Surrah AS, Al-Sa'doni HH, Abdalla MY. Palladium-based chemotherapeutic agents: Routes toward complexes with good antitumor activity. Cancer Therapy. 2008;**6**:1-10

[56] Niehoff A-C, Moosmann A, Söbbing J, Wiehe A, Mulac D, Wehe CA, et al. A palladium label to monitor nanoparticle-assisted drug delivery of a photosensitizer into tumor spheroids by elemental bioimaging. Metallomics. 2014;**6**(1):77-81

[57] Dumas A, Couvreur P. Palladium: A future key player in the nanomedical field? Chemical Science. 2015;**6**(4):2153-2157

[58] Adams CP, Walker KA, Obare SO, Docherty KM. Size-dependent antimicrobial effects of novel palladium nanoparticles. PLoS One. 2014;**9**(1):e85981

[59] McGrath AJ, Chien Y-H, Cheong S, Herman DA, Watt J, Henning AM, et al. Gold over branched palladium nanostructures for photothermal cancer therapy. ACS Nano. 2015;**9**(12):12283-12291

[60] Rezaei B, Shams-Ghahfarokhi L, Havakeshian E, Ensafi AA. An electrochemical biosensor based on nanoporous stainless steel modified by gold and palladium nanoparticles for simultaneous determination of levodopa and uric acid. Talanta. 2016;**158**:42-50

[61] Baccar H, Ktari T, Abdelghani A. Functionalized palladium nanoparticles for hydrogen peroxide biosensor. International Journal of Electrochemistry. 2011

[62] Han HS, You J-M, Jeong H, Jeon S. Synthesis of graphene oxide grafted poly (lactic acid) with palladium nanoparticles and its application to serotonin sensing. Applied Surface Science. 2013;**284**:438-445

[63] Laurent S, Forge D, Port M, Roch A, Robic C, Vander Elst L, et al. Magnetic iron oxide nanoparticles: Synthesis, stabilization, vectorization, physicochemical characterizations, and biological applications. Chemical Reviews. 2008;**108**(6):2064-2110

[64] Rahman MM, Khan SB, Jamal A, Faisal M, Aisiri AM. Iron oxide nanoparticles. In: Nanomaterials. InTech; 2011

[65] Khan SA, Gambhir S, Ahmad A. Extracellular biosynthesis of gadolinium oxide (Gd_2O_3) nanoparticles, their biodistribution and bioconjugation with the chemically modified anticancer drug taxol. Beilstein Journal of Nanotechnology. 2014;**5**:249

[66] Mahmoudi M, Hosseinkhani H, Hosseinkhani M, Boutry S, Simchi A, Journeay WS, et al. Magnetic resonance imaging tracking of stem cells in vivo using iron oxide nanoparticles as a tool for the advancement of clinical regenerative medicine. Chemical Reviews. 2010;**111**(2):253-280

[67] Laurent S, Saei AA, Behzadi S, Panahifar A, Mahmoudi M. Superparamagnetic iron oxide nanoparticles for delivery of therapeutic agents: Opportunities and challenges. Expert Opinion on Drug Delivery. 2014;**11**(9):1449-1470

[68] De Toledo LAS, Rosseto HC, Bruschi ML. Iron oxide magnetic nanoparticles as antimicrobials for therapeutics. Pharmaceutical Development and Technology. 2017:1-30 (just-accepted)

[69] Bakhtiary Z, Saei AA, Hajipour MJ, Raoufi M, Vermesh O, Mahmoudi M. Targeted superparamagnetic iron oxide nanoparticles for early detection of cancer: Possibilities and challenges. Nanomedicine: Nanotechnology, Biology and Medicine. 2016;**12**(2):287-307

[70] Sun C, Lee JS, Zhang M. Magnetic nanoparticles in MR imaging and drug delivery. Advanced Drug Delivery Reviews. 2008;**60**(11):1252-1265

[71] Rauch J, Kolch W, Laurent S, Mahmoudi M. Big signals from small particles: Regulation of cell signaling pathways by nanoparticles. Chemical Reviews. 2013;**113**(5):3391-3406

[72] Hola K, Markova Z, Zoppellaro G, Tucek J, Zboril R. Tailored functionalization of iron oxide nanoparticles for MRI, drug delivery, magnetic separation and immobilization of biosubstances. Biotechnology Advances. 2015;**33**(6):1162-1176

[73] Mahdy SA, Raheed QJ, Kalaichelvan P. Antimicrobial activity of zero-valent iron nanoparticles. International Journal of Modern Engineering Research. 2012;**2**(1):578-581

[74] Chekina N, Horák D, Jendelová P, Trchová M, Beneš MJ, Hrubý M, et al. Fluorescent magnetic nanoparticles for biomedical applications. Journal of Materials Chemistry. 2011;**21**(21):7630-7639

[75] Narendhar C, Anbarasu S, Divakar S, Gunaseelan R, Sundaram V, Gopu G, et al. Antimicrobial activity of chitosan coated iron oxide nanoparticles. International Journal of ChemTech Research. 2014;**6**(3):2210-2212

[76] Espinosa A, Di Corato R, Kolosnjaj-Tabi J, Flaud P, Pellegrino T, Wilhelm C. Duality of iron oxide nanoparticles in cancer therapy: Amplification of heating efficiency

by magnetic hyperthermia and photothermal bimodal treatment. ACS Nano. 2016;**10**(2):2436-2446

[77] Peterson RD, Cunningham BT, Andrade JE. A photonic crystal biosensor assay for ferritin utilizing iron-oxide nanoparticles. Biosensors and Bioelectronics. 2014;**56**:320-327

[78] Santana SDF. Magnetic Nanoparticles for Biocatalysis and Bioseparation. Portugal: Faculdade de Ciências e Tecnologia; 2011. Available from: http://hdl.handle.net/10362/6259

[79] Chen H, Burnett J, Zhang F, Zhang J, Paholak H, Sun D. Highly crystallized iron oxide nanoparticles as effective and biodegradable mediators for photothermal cancer therapy. Journal of Materials Chemistry B. 2014;**2**(7):757-765

[80] Zhang Y, R Nayak T, Hong H, Cai W. Biomedical applications of zinc oxide nanomaterials. Current Molecular Medicine. 2013;**13**(10):1633-1645

[81] Wang ZL. Zinc oxide nanostructures: Growth, properties and applications. Journal of Physics: Condensed Matter. 2004;**16**(25):R829

[82] Rasmussen JW, Martinez E, Louka P, Wingett DG. Zinc oxide nanoparticles for selective destruction of tumor cells and potential for drug delivery applications. Expert Opinion on Drug Delivery. 2010;**7**(9):1063-1077

[83] Gulia S, Kakkar R. ZnO quantum dots for biomedical applications. Advanced Materials Letters. 2013;**4**(12):876-887

[84] Sudhagar S, Sathya S, Pandian K, Lakshmi BS. Targeting and sensing cancer cells with ZnO nanoprobes in vitro. Biotechnology Letters. 2011;**33**(9):1891-1896

[85] Tripathy N, Ahmad R, Ko HA, Khang G, Hahn Y-B. Enhanced anticancer potency using an acid-responsive ZnO-incorporated liposomal drug-delivery system. Nanoscale. 2015;**7**(9):4088-4096

[86] Xiong HM. ZnO nanoparticles applied to bioimaging and drug delivery. Advanced Materials. 2013;**25**(37):5329-5335

[87] Zhang H, Chen B, Jiang H, Wang C, Wang H, Wang X. A strategy for ZnO nanorod mediated multi-mode cancer treatment. Biomaterials. 2011;**32**(7):1906-1914

[88] Yang S-C, Shen Y-C, Lu T-C, Yang T-L, Huang J-J. Tumor detection strategy using ZnO light-emitting nanoprobes. Nanotechnology. 2012;**23**(5):055202

[89] Major JL, Parigi G, Luchinat C, Meade TJ. The synthesis and in vitro testing of a zinc-activated MRI contrast agent. Proceedings of the National Academy of Sciences. 2007;**104**(35):13881-13886

[90] Meruvu H, Vangalapati M, Chippada SC, Bammidi SR. Synthesis and characterization of zinc oxide nanoparticles and its antimicrobial activity against *Bacillus subtilis* and *Escherichia coli*. Rasayan Journal of Chemistry. 2011;**4**(1):217-222

[91] Inbasekaran S, Senthil R, Ramamurthy G, Sastry T. Biosensor using zinc oxide nanoparticles. International Journal of Innovative Research in Science, Engineering and Technology. 2014;**3**(1):8601-8606

[92] Lei Y, Luo N, Yan X, Zhao Y, Zhang G, Zhang Y. A highly sensitive electrochemical biosensor based on zinc oxide nanotetrapods for L-lactic acid detection. Nanoscale. 2012;**4**(11):3438-3443

[93] Wei A, Pan L, Huang W. Recent progress in the ZnO nanostructure-based sensors. Materials Science and Engineering: B. 2011;**176**(18):1409-1421

[94] Wu Y, Fu S, Tok A, Zeng X, Lim C, Kwek L, et al. A dual-colored bio-marker made of doped ZnO nanocrystals. Nanotechnology. 2008;**19**(34):345605

[95] Ma J, Zhu W, Tian Y, Wang Z. Preparation of zinc oxide-starch nanocomposite and its application on coating. Nanoscale Research Letters. 2016;**11**(1):200

[96] Lu X, Lv X, Sun Z, Zheng Y. Nanocomposites of poly (L-lactide) and surface-grafted TiO_2 nanoparticles: Synthesis and characterization. European Polymer Journal. 2008;**44**(8):2476-2481

[97] Xu W, Xie W, Huang X, Chen X, Huang N, Wang X, et al. The graphene oxide and chitosan biopolymer loads TiO_2 for antibacterial and preservative research. Food Chemistry. 2017;**221**:267-277

[98] Rahman IA, Padavettan V. Synthesis of silica nanoparticles by sol-gel: Size-dependent properties, surface modification, and applications in silica-polymer nanocomposites—A review. Journal of Nanomaterials. 2012;**2012**:8

[99] Gardella L, Basso A, Prato M, Monticelli O. PLA/POSS nanofibers: A novel system for the immobilization of metal nanoparticles. ACS Applied Materials & Interfaces. 2013;**5**(16):7688-7692

[100] Nabipour Chakoli A, Wan J, Feng JT, Amirian M, Sui JH, Cai W. Functionalization of multiwalled carbon nanotubes for reinforcing of poly(l-lactide-co-ε-caprolactone) biodegradable copolymers. Applied Surface Science. 2009;**256**:170-177

[101] Dai H. Carbon nanotubes: Synthesis, integration, and properties. Accounts of Chemical Research. 2002;**35**(12):1035-1044

[102] Kong H, Gao C, Yan D. Controlled functionalization of multiwalled carbon nanotubes by in situ atom transfer radical polymerization. Journal of the American Chemical Society. 2004;**126**(2):412-413

[103] Qin S, Qin D, Ford WT, Herrera JE, Resasco DE, Bachilo SM, et al. Solubilization and purification of single-wall carbon nanotubes in water by in situ radical polymerization of sodium 4-styrenesulfonate. Macromolecules. 2004;**37**(11):3965-3967

[104] Kim H-S, Park BH, Yoon J-S, Jin H-J. Thermal and electrical properties of poly (L-lactide)-graft-multiwalled carbon nanotube composites. European Polymer Journal. 2007;**43**(5):1729-1735

[105] Chen GX, Kim HS, Park BH, Yoon JS. Synthesis of poly(L-lactide)-functionalized multiwalled carbon nanotubes by ring-opening polymerization. Macromolecular Chemistry and Physics. 2007;**208**(4):389-398

[106] Saeed K, Park S-Y, Lee H-J, Baek J-B, Huh W-S. Preparation of electrospun nanofibers of carbon nanotube/polycaprolactone nanocomposite. Polymer. 2006;**47**(23):8019-8025

[107] Chen G-X, Shimizu H. Multiwalled carbon nanotubes grafted with polyhedral oligomeric silsesquioxane and its dispersion in poly(L-lactide) matrix. Polymer. 2008;**49**(4):943-951

[108] Tasis D, Tagmatarchis N, Bianco A, Prato M. Chemistry of carbon nanotubes. Chemical Reviews. 2006;**106**(3):1105-1136

[109] Balasubramanian K, Burghard M. Chemically functionalized carbon nanotubes. Small. 2005;**1**(2):180-192

[110] Chen J, Hamon MA, Hu H, Chen Y, Rao AM, Eklund PC, et al. Solution properties of single-walled carbon nanotubes. Science. 1998;**282**(5386):95-98

[111] Gu Z, Peng H, Hauge R, Smalley R, Margrave J. Cutting single-wall carbon nanotubes through fluorination. Nano Letters. 2002;**2**(9):1009-1013

[112] Banerjee S, Hemraj-Benny T, Wong SS. Covalent surface chemistry of single-walled carbon nanotubes. Advanced Materials. 2005;**17**(1):17-29

[113] Rinzler A, Liu J, Dai H, Nikolaev P, Huffman C, Rodriguez-Macias F, et al. Large-scale purification of single-wall carbon nanotubes: Process, product, and characterization. Applied Physics A: Materials Science & Processing. 1998;**67**(1):29-37

[114] Dujardin E, Ebbesen TW, Krishnan A, Treacy MM. Purification of single-shell nanotubes. Advanced Materials. 1998;**10**(8):611-613

[115] Wong SS, Joselevich E, Woolley AT, Cheung CL, Lieber CM. Covalently functionalized nanotubes as nanometre-sized probes in chemistry and biology. Nature. 1998;**394**(6688):52

[116] Moore VC, Strano MS, Haroz EH, Hauge RH, Smalley RE, Schmidt J, et al. Individually suspended single-walled carbon nanotubes in various surfactants. Nano Letters. 2003;**3**(10):1379-1382

[117] Bianco A, Kostarelos K, Partidos CD, Prato M. Biomedical applications of functionalised carbon nanotubes. Chemical Communications. 2005;(5):571-577

[118] Coleman JN, Cadek M, Ryan KP, Fonseca A, Nagy JB, Blau WJ, et al. Reinforcement of polymers with carbon nanotubes. The role of an ordered polymer interfacial region. Experiment and modeling. Polymer. 2006;**47**(26):8556-8561

[119] Zheng M, Jagota A, Semke ED, Diner BA, Mclean RS, Lustig SR, et al. DNA-assisted dispersion and separation of carbon nanotubes. Nature Materials. 2003;**2**(5):338

[120] Sahithi K, Swetha M, Ramasamy K, Srinivasan N, Selvamurugan N. Polymeric composites containing carbon nanotubes for bone tissue engineering. International Journal of Biological Macromolecules. 2010;**46**(3):281-283

[121] Wang Y, Li Z, Wang J, Li J, Lin Y. Graphene and graphene oxide: Biofunctionalization and applications in biotechnology. Trends in Biotechnology. 2011;**29**(5):205-212

[122] Geng L-H, Peng X-F, Jing X, Li L-W, Huang A, Xu B-P, et al. Investigation of poly(L-lactic acid)/graphene oxide composites crystallization and nanopore foaming behaviors via supercritical carbon dioxide low temperature foaming. Journal of Materials Research. 2016;**31**(3):348-359

[123] Venkatesan J, Kim S-K. Nano-hydroxyapatite composite biomaterials for bone tissue engineering—A review. Journal of Biomedical Nanotechnology. 2014;**10**(10):3124-3140

[124] Xu X, Chen X, Liu A, Hong Z, Jing X. Electrospun poly(L-lactide)-grafted hydroxyapatite/poly(L-lactide) nanocomposite fibers. European Polymer Journal. 2007;**43**(8):3187-3196

[125] Arruebo M, Fernández-Pacheco R, Ibarra MR, Santamaría J. Magnetic nanoparticles for drug delivery. Nano Today. 2007;**2**(3):22-32

[126] Dobson J. Magnetic nanoparticles for drug delivery. Drug Development Research. 2006;**67**(1):55-60

[127] Wan J, Cai W, Feng J, Meng X, Liu E. In situ decoration of carbon nanotubes with nearly monodisperse magnetite nanoparticles in liquid polyols. Journal of Materials Chemistry. 2007;**17**(12):1188-1192

[128] Feng J, Cai W, Sui J, Li Z, Wan J, Chakoli AN. Poly(L-lactide) brushes on magnetic multiwalled carbon nanotubes by in-situ ring-opening polymerization. Polymer. 2008;**49**(23):4989-4994